THE WILLIAM LOWELL PUTNAM MATHEMATICAL COMPETITION

PROBLEMS AND SOLUTIONS: 1965–1984

Gerald L. Alexanderson
University of Santa Clara

Leonard F. Klosinski
University of Santa Clara

Loren C. Larson
St. Olaf College

The WILLIAM LOWELL PUTNAM MATHEMATICAL COMPETITION

PROBLEMS AND SOLUTIONS: 1965–1984

Edited by

Gerald L. Alexanderson
Leonard F. Klosinski
Loren C. Larson

Published and distributed by

The Mathematical Association of America

Library of Congress Catalog Card Number 85-062263

ISBN 0-88385-441-4

Printed in the United States of America

Current printing (last digit):

10 9 8 7 6 5 4 3 2 1

DEDICATED TO

THE PUTNAM CONTESTANTS

PREFACE

Let us make clear from the start that we have not tried with this collection to imitate the scholarly and extensive treatment of the first twenty-five contests by Gleason, Greenwood, and Kelly (*The William Lowell Putnam Mathematical Competition / Problems and Solutions: 1938–1964*. Washington: MAA, 1980). That splendid volume shows the years of work spent in following up on problems, compiling better solutions, and tracing effects of some of the problems in subsequent work.

We have done none of that here. We have compiled material essentially already available in the *American Mathematical Monthly* and *Mathematics Magazine*, correcting in several cases solutions where errors had crept in. The present volume is mainly an attempt to put together in convenient form existing material. A volume comparable to the Gleason, Greenwood, Kelly book will have to wait for another time. We hope that in the meantime the present collection will benefit students interested in preparing for the Competition, faculty who wish to organize problem seminars, or any others just interested in problems.

For information about the history of the Putnam Competition, we refer the reader to the excellent articles by Garrett Birkhoff and L. E. Bush in the earlier collection. These articles also appeared in the *Monthly* in 1965. We are happy to have in the present collection a further bit of information about the origins of the Competition, an essay on the first contest by Herbert Robbins as told to Alan Tucker.

We have summarized lists of winning teams and individual participants; more extensive information on winners and teams appears in annual reports in the *Monthly*.

Our work would have been much more difficult had we not had the reports of the Competition carefully prepared by former directors of the Competition, James H. McKay (Oakland University) and Abraham P. Hillman (University of New Mexico). We wish especially to thank them for their many contributions over the years and specifically for their excellent reports. They are largely responsible for the presentation of solutions that have appeared in the *Monthly* during their directorships, though, of course, they had the benefit of having the solutions given them by members of the Questions Committees over those years. And, of course, had the members of the Questions Committee not provided the questions (and in many cases solutions) there would have been no Competition. We therefore wish to thank the members of the Questions Committee: H. S. M. Coxeter (University of Toronto), Adriano M. Garsia (California Institute of Technology), Robert E. Greenwood (University of Texas, Austin), Nicholas D. Kazarinoff (University of Michigan, Ann Arbor), Leo Moser (University of Alberta), Albert Wilansky (Lehigh University), Warren S. Loud (University of Minnesota, Minneapolis), Murray S. Klamkin (Ford Scientific Laboratories), Nathan S. Mendelsohn (University of Manitoba), Donald J. Newman (Yeshiva University), J. Ian Richards (University of Minnesota, Minneapolis), Gulbank D. Chakerian (University of California, Davis), Joseph D. E. Konhauser (Macalester College), Richard J. Bumby (Rutgers University, New Brunswick), Lawrence A. Zalcman (University of Maryland, College Park), Edward J. Barbeau, Jr. (University of Toronto), Kenneth B. Stolarsky (University of Illinois, Urbana-Champaign), Joel H. Spencer (State University of New York, Stony Brook), William J. Firey (Oregon State University), Douglas A. Hensley (Texas A & M University),

Melvin Hochster (University of Michigan, Ann Arbor), Bruce A. Reznick (University of Illinois, Urbana-Champaign), and Richard P. Stanley (Massachusetts Institute of Technology).

We would further like to thank Alan Tucker, Chairman of the Publications Committee of the MAA, A. B. Willcox, Executive Director, and Beverly Joy Ruedi of the Editorial Office of the MAA.

Gerald L. Alexanderson
Leonard F. Klosinski
Loren C. Larson

March, 1985

CONTENTS

RECOLLECTIONS OF THE FIRST PUTNAM EXAMINATION

HERBERT ROBBINS
as told to Alan Tucker

The well-known story of the origin of the initial 1933 Putnam contest in mathematics is as follows (I believe this story to be mostly true). During half-time of the 1931 Harvard-Army football game, President A. Lawrence Lowell said to the Commandant of the U.S. Military Academy that while Army was showing that it could trounce Harvard in football, Harvard would just as easily win any contest of a more academic nature. The Commandant took President Lowell up on his challenge and it was decided to have a mathematics contest between the two schools. I would guess that the field of mathematics was chosen because it is a subject that was studied at both West Point and Harvard (all cadets, then as now, took two years of math) and because a relative of President Lowell, George Putnam, was an amateur mathematician who was involved in arrangements for the contest and got it named after his relative William Lowell Putnam.

I came to Harvard in the fall of 1931 with what I thought was some knowledge of the humanities, but no mathematics beyond quadratic equations. To address this deficiency, I enrolled in Math A, Analytic Geometry and Calculus. The texts were *Analytic Geometry* by Osgood and Graustein and *Calculus* by Osgood. These books were casual in their treatment of real numbers and limits but had challenging problems that assumed a good knowledge of physics. The best-prepared students in Math A were put in the section taught by the eminent Professor Julian Lowell Coolidge, and the rest were taught by junior instructors. I was put in the section taught by Sumner B. Myers. I skipped a good many of the classes (being occupied with certain extracurricular interests most of the year) but did well in the tests.

At the end of the course in May, 1932, I was invited to be part of the team that would represent Harvard in the mathematics competition with West Point the following spring. Since cadets took two years of mathematics, the Harvard team was restricted to students who would be completing their second year of mathematics at the time of the test. Selection for this team led me to continue my study of mathematics for a second year.

The Harvard Mathematics Department assigned Professor Marston Morse to coach the team, and we met with him about four times during the fall and winter. It was assumed that our Harvard intellects would easily carry the day, and our meetings with Morse were spent in general conversation rather than problem-solving. However, these sessions were very important to me, for I became impressed by Morse as a person and by the incomprehensibility of the mathematics he spoke of.

I had many conversations with Morse in addition to the Putnam sessions. Morse was in a low mood then because his wife had recently left him to marry Professor William Fogg Osgood, a distinguished Harvard mathematician with a long white beard who was 28 years Morse's senior (this scandal forced Osgood to leave Harvard). However, my association with Morse was not enough to persuade me to pick mathematics as my major just yet; I was experimenting then with majors in the sciences and philosophy.

One spring weekend the Harvard team traveled to West Point for the competition. There was a morning and afternoon session, and the problems were rather cut and dried, technical integrations and the like, with little call for originality. The highlight of the weekend for me was a date Saturday night in New York City with a girl I had met the previous summer.

Back at Harvard we found out, to our shame, that we had lost the competition to Army. I was told that I had done well on the exam (we never saw our exam books) and decided to be a math major. I never would have studied more than a year of mathematics, much less have become for a time a mathematician, were it not for my experience with the Putnam competition.

The next year Morse left Harvard to go to the Institute for Advanced Study. He told me to continue my studies, to get a PhD in math at Harvard, and then to get in touch with him. I had no further contact with Morse until five years later when I defended my thesis at Harvard and sent him a telegram: "Have PhD in mathematics." His response was equally brief: "You are my assistant starting September 1."

It would seem that the reason I finally became a math major was that most Harvard mathematics professors were rather pompous know-it-alls and that I wanted to show them that any reasonably bright person could do mathematics. Unfortunately, I won the battle but lost the war.

Herbert Robbins was the subject of an interview in the January 1984 *College Mathematics Journal.* He was co-author with Richard Courant of *What is Mathematics*? and is Higgins Professor of Mathematical Statistics at Columbia University.

PROBLEMS

THE TWENTY-SIXTH WILLIAM LOWELL PUTNAM MATHEMATICAL COMPETITION

November 20, 1965

A-1. Let ABC be a triangle with angle $A <$angle $C < 90° <$angle B. Consider the bisectors of the external angles at A and B, each measured from the vertex to the opposite side (extended). Suppose both of these line-segments are equal to AB. Compute the angle A.

A-2. Show that, for any positive integer n,

$$\sum_{r=0}^{[(n-1)/2]} \left\{\frac{n-2r}{n}\binom{n}{r}\right\}^2 = \frac{1}{n}\binom{2n-2}{n-1},$$

where $[x]$ means the greatest integer not exceeding x, and $\binom{n}{r}$ is the binomial coefficient "n choose r," with the convention $\binom{n}{0}=1$.

A-3. Show that, for any sequence $a_1, a_2, \cdots$ of real numbers, the two conditions

(A)
$$\lim_{n\to\infty} \frac{e^{(ia_1)} + e^{(ia_2)} + \cdots + e^{(ia_n)}}{n} = \alpha$$

and

(B)
$$\lim_{n\to\infty} \frac{e^{(ia_1)} + e^{(ia_4)} + \cdots + e^{(ia_{n^2})}}{n^2} = \alpha$$

are equivalent.

A-4. At a party, assume that no boy dances with every girl but each girl dances with at least one boy. Prove that there are two couples gb and $g'b'$ which dance whereas b does not dance with g' nor does g dance with b'.

A-5. In how many ways can the integers from 1 to n be ordered subject to the condition that, except for the first integer on the left, every integer differs by 1 from some integer to the left of it?

A-6. In the plane with orthogonal Cartesian coordinates x and y, prove that the line whose equation is $ux+vy=1$ will be tangent to the curve $x^m+y^m=1$ (where $m>1$) if and only if $u^n+v^n=1$ and $m^{-1}+n^{-1}=1$.

B-1. Evaluate

$$\lim_{n\to\infty} \int_0^1 \int_0^1 \cdots \int_0^1 \cos^2\left\{\frac{\pi}{2n}(x_1 + x_2 + \cdots x_n)\right\} dx_1 dx_2 \cdots dx_n.$$

B-2. In a round-robin tournament with n players $P_1, P_2, \cdots, P_n$ (where $n>1$), each player plays one game with each of the other players and the rules are such that no ties can occur. Let ω_r and l_r be the number of games won and lost, respectively, by P_r. Show that

$$\sum_{r=1}^{n} w_r^2 = \sum_{r=1}^{n} l_r^2.$$

B-3. Prove that there are exactly three right-angled triangles whose sides are integers while the area is numerically equal to twice the perimeter.

B-4. Consider the function

$$f(x, n) = \frac{\binom{n}{0} + \binom{n}{2} x + \binom{n}{4} x^2 + \cdots}{\binom{n}{1} + \binom{n}{3} x + \binom{n}{5} x^2 + \cdots},$$

where n is a positive integer. Express $f(x, n+1)$ rationally in terms of $f(x, n)$ and x. Hence, or otherwise, evaluate $\lim_{n\to\infty} f(x, n)$ for suitable fixed values of x. (The symbols $\binom{n}{r}$ represent the binomial coefficients.)

B-5. Consider collections of unordered pairs of V different objects $a, b, c, \cdots, k$. Three pairs such as bc, ca, ab are said to form a triangle. Prove that, if $4E \leqq V^2$, it is possible to choose E pairs so that no triangle is formed.

B-6. If A, B, C, D are four distinct points such that every circle through A and B intersects (or coincides with) every circle through C and D, prove that the four points are either collinear (all of one line) or concyclic (all on one circle).

THE TWENTY-SEVENTH WILLIAM LOWELL PUTNAM MATHEMATICAL COMPETITION

November 19, 1966

A-1. Let $f(n)$ be the sum of the first n terms of the sequence $0, 1, 1, 2, 2, 3, 3, 4, \cdots$, where the nth term is given by

$$a_n = \begin{cases} n/2 & \text{if } n \text{ is even,} \\ (n-1)/2 & \text{if } n \text{ is odd.} \end{cases}$$

Show that if x and y are positive integers and $x > y$ then $xy = f(x+y) - f(x-y)$.

A-2. Let a, b, c be the lengths of the sides of a triangle, let $p = (a+b+c)/2$, and r be the radius of the inscribed cricle. Show that

$$\frac{1}{(p-a)^2} + \frac{1}{(p-b)^2} + \frac{1}{(p-c)^2} \geqq \frac{1}{r^2}.$$

A-3. Let $0 < x_1 < 1$ and $x_{n+1} = x_n(1-x_n)$, $n = 1, 2, 3, \cdots$. Show that

$$\lim_{n\to\infty} nx_n = 1.$$

A-4. Prove that after deleting the perfect squares from the list of positive integers the number we find in the nth position is equal to $n + \{\sqrt{n}\}$, where $\{\sqrt{n}\}$ denotes the integer closest to $\sqrt{n}$.

A-5. Let C denote the family of continuous functions on the real axis. Let T be a mapping of C into C which has the following properties:

1. T *is linear*, i.e. $T(c_1\psi_1 + c_2\psi_2) = c_1T\psi_1 + c_2T\psi_2$, for c_1 and c_2 real and ψ_1 and ψ_2 in C.

2. T *is local*, i.e. if $\psi_1 \equiv \psi_2$ in some interval I then also $T\psi_1 \equiv T\psi_2$ holds in I.

Show that T must necessarily be of the form $T\psi(x) = f(x)\psi(x)$ where $f(x)$ is a suitable continuous function.

A-6. Justify the statement that

$$3 = \sqrt{1 + 2\sqrt{1 + 3\sqrt{1 + 4\sqrt{1 + 5\sqrt{1 + \cdots}}}}}.$$

B-1. Let a convex polygon P be contained in a square of side one. Show that the sum of the squares of the sides of P is less than or equal to 4.

B-2. Prove that among any ten consecutive integers at least one is relatively prime to each of the others.

B-3. Show that if the series

$$\sum_{n=1}^{\infty} \frac{1}{p_n}$$

is convergent, where $p_1, p_2, p_3, \cdots, p_n, \cdots$ are positive real numbers, then the series

$$\sum_{n=1}^{\infty} \frac{n^2}{(p_1 + p_2 + \cdots + p_n)^2} p_n$$

is also convergent.

B-4. Let $0<a_1<a_2<\cdots<a_{mn+1}$ be $mn+1$ integers. Prove that you can select either $m+1$ of them no one of which divides any other, or $n+1$ of them each dividing the following one.

B-5. Given $n(\geqq 3)$ distinct points in the plane, no three of which are on the same straight line, prove that there exists a simple closed polygon with these points as vertices.

B-6. Show that all solutions of the differential equation $y''+e^x y=0$ remain bounded as $x \to \infty$.

THE TWENTY-EIGHTH WILLIAM LOWELL PUTNAM MATHEMATICAL COMPETITION

December 2, 1967

A-1. Let $f(x)=a_1\sin x+a_2\sin 2x+\cdots+a_n\sin nx$, where $a_1, a_2, \cdots, a_n$ are real numbers and where n is a positive integer. Given that $|f(x)|\leqq|\sin x|$ for all real x, prove that

$$|a_1+2a_2+\cdots+na_n|\leqq 1.$$

A-2. Define S_0 to be 1. For $n\geqq 1$, let S_n be the number of $n\times n$ matrices whose elements are nonnegative integers with the property that $a_{ij}=a_{ji}$, $(i,j=1, 2, \cdots, n)$ and where $\sum_{i=1}^{n} a_{ij}=1$, $(j=1, 2, \cdots, n)$. Prove

$$\text{(a) } S_{n+1}=S_n+nS_{n-1},$$

$$\text{(b) } \sum_{n=0}^{\infty} S_n \frac{x^n}{n!}=\exp(x+x^2/2), \qquad \text{where } \exp(x)=e^x.$$

A-3. Consider polynomial forms ax^2-bx+c with integer coefficients which have two distinct zeros in the open interval $0<x<1$. Exhibit with a proof the least positive integer value of a for which such a polynomial exists.

A-4. Show that if $\lambda>\frac{1}{2}$ there does not exist a real-valued function u such that for all x in the closed interval $0\leqq x\leqq 1$, $u(x)=1+\lambda\int_x^1 u(y)\,u(y-x)\,dy$.

A-5. Show that in a convex region in the plane whose boundary contains at most a finite number of straight line segments and whose area is greater than $\pi/4$ there is at least one pair of points a unit distance apart.

A-6. Given real numbers $\{a_i\}$ and $\{b_i\}$, $(i=1, 2, 3, 4)$, such that $a_1b_2-a_2b_1\neq 0$. Consider the set of all solutions (x_1, x_2, x_3, x_4) of the simultaneous equations

$$a_1x_1+a_2x_2+a_3x_3+a_4x_4=0 \text{ and } b_1x_1+b_2x_2+b_3x_3+b_4x_4=0,$$

for which no x_i $(i=1, 2, 3, 4)$ is zero. Each such solution generates a 4-tuple of plus and minus signs (signum x_1, signum x_2, signum x_3, signum x_4).

(a) Determine, with a proof, the maximum number of distinct 4-tuples possible.

(b) Investigate necessary and sufficient conditions on the real numbers $\{a_i\}$ and $\{b_i\}$ such that the above maximum number of 4-tuples is attained.

B-1. Let $(ABCDEF)$ be a hexagon inscribed in a circle of radius r. Show that if $\overline{AB}=\overline{CD}=\overline{EF}=r$, then the midpoints of $\overline{BC}$, $\overline{DE}$, $\overline{FA}$ are the vertices of an equilateral triangle.

B-2. Let $0\leqq p\leqq 1$ and $0\leqq r\leqq 1$ and consider the identities

$$\text{(a) } (px+(1-p)y)^2=Ax^2+Bxy+Cy^2,$$

$$\text{(b) } (px+(1-p)y)(rx+(1-r)y)=\alpha x^2+\beta xy+\gamma y^2.$$

Show that (with respect to p and r)

$$\text{(a) } \max\{A, B, C\}\geqq 4/9,$$

$$\text{(b) } \max\{\alpha, \beta, \gamma\}\geqq 4/9.$$

B-3. If f and g are continuous and periodic functions with period 1 on the real line, then $\lim_{n\to\infty} \int_0^1 f(x)\, g(nx)\, dx = (\int_0^1 f(x)dx)(\int_0^1 g(x)dx)$.

B-4. (a) A certain locker room contains n lockers numbered $1, 2, 3, \cdots, n$ and all are originally locked. An attendant performs a sequence of operations $T_1, T_2, \cdots, T_n$ whereby with the operation T_k, $1 \leqq k \leqq n$, the condition of being locked or unlocked is changed for all those lockers and only those lockers whose numbers are multiples of k. After all the n operations have been performed it is observed that all lockers whose numbers are perfect squares (and only those lockers) are now open or unlocked. Prove this mathematically.

(b) Investigate in a meaningful mathematical way a procedure or set of operations similar to those above which will produce the set of cubes, or the set of numbers of the form $2m^2$, or the set of numbers of the form m^2+1, or some nontrivial similar set of your own selection.

B-5. Show that the sum of the first n terms in the binomial expansion of $(2-1)^{-n}$ is $\frac{1}{2}$, where n is a positive integer.

B-6. Let f be a real-valued function having partial derivatives and which is defined for $x^2+y^2 \leqq 1$ and is such that $|f(x, y)| \leqq 1$. Show that there exists a point (x_0, y_0) in the interior of the unit circle such that

$$\left(\frac{\partial f}{\partial x}(x_0, y_0)\right)^2 + \left(\frac{\partial f}{\partial y}(x_0, y_0)\right)^2 \leqq 16.$$

THE TWENTY-NINTH WILLIAM LOWELL PUTNAM MATHEMATICAL COMPETITION

December 7, 1968

A-1. Prove

$$\frac{22}{7} - \pi = \int_0^1 \frac{x^4(1-x)^4}{1+x^2}\,dx.$$

A-2. Given integers a, b, e, c, d, and f with $ad \neq bc$, and given a real number $\epsilon > 0$, show that there exist rational numbers r and s for which

$$0 < |ra + sb - e| < \epsilon,$$
$$0 < |rc + sd - f| < \epsilon.$$

A-3. Prove that a list can be made of all the subsets of a finite set in such a way that (i) the empty set is first in the list, (ii) each subset occurs exactly once, (iii) each subset in the list is obtained either by adding one element to the preceding subset or by deleting one element of the preceding subset.

A-4. Given n points on the sphere $\{(x, y, z): x^2+y^2+z^2=1\}$, demonstrate that the sum of the squares of the distances between them does not exceed n^2.

A-5. Let V be the collection of all quadratic polynomials P with real coefficients such that $|P(x)| \leqq 1$ for all x on the closed interval $[0, 1]$. Determine

$$\sup\{|P'(0)| : P \in V\}.$$

A-6. Determine all polynomials of the form $\sum_0^n a_i x^{n-i}$ with $a_i = \pm 1$ $(0 \leqq i \leqq n,\ 1 \leqq n < \infty)$ such that each has only real zeros.

B-1. The temperatures in Chicago and Detroit are x° and y°, respectively. These temperatures are not assumed to be independent; namely, we are given:
(i) $P(x^\circ = 70^\circ)$, the probability that the temperature in Chicago is 70°,
(ii) $P(y^\circ = 70^\circ)$, and
(iii) $P(\max(x^\circ, y^\circ) = 70^\circ)$.
Determine $P(\min(x^\circ, y^\circ) = 70^\circ)$.

B-2. A is a subset of a finite group G (with group operation called multiplication), and A contains more than one half of the elements of G. Prove that each element of G is the product of two elements of A.

B-3. Assume that a 60° angle cannot be trisected with ruler and compass alone. Prove that if n is a positive multiple of 3, then no angle of $360/n$ degrees can be trisected with ruler and compass alone.

B-4. Show that if f is real-valued and continuous on $(-\infty, \infty)$ and $\int_{-\infty}^{\infty} f(x)dx$ exists, then

$$\int_{-\infty}^{\infty} f\left(x - \frac{1}{x}\right) dx = \int_{-\infty}^{\infty} f(x)dx.$$

B-5. Let p be a prime number. Let J be the set of all 2×2 matrices $\binom{ab}{cd}$ whose entries are chosen from $\{0, 1, 2, \cdots, p-1\}$ and satisfy the conditions $a+d \equiv 1 \pmod p$, $ad - bc \equiv 0 \pmod p$.

Determine how many members J has.

B-6. A set of real numbers is called compact if it is closed and bounded. Show that there does not exist a sequence $\{K_n\}_{n=0}^{\infty}$ of compact sets of rational numbers such that each compact set of rationals is contained in at least one K_n.

THE THIRTIETH WILLIAM LOWELL PUTNAM MATHEMATICAL COMPETITION

December 6, 1969

A-1. Let $f(x, y)$ be a polynomial with real coefficients in the real variables x and y defined over the entire x-y plane. What are the possibilities for the range of $f(x, y)$?

A-2. Let D_n be the determinant of order n of which the element in the ith row and the jth column is the absolute value of the difference of i and j. Show that D_n is equal to

$$(-1)^{n-1}(n-1)2^{n-2}.$$

A-3. Let P be a non-self-intersecting closed polygon with n sides. Let its vertices be $P_1, P_2, \cdots, P_n$. Let m other points, $Q_1, Q_2, \cdots, Q_m$ interior to P be given. Let the figure be triangulated. This means that certain pairs of the $(n+m)$ points $P_1, \cdots, Q_m$ are connected by line segments such that (i) the resulting figure consists exclusively of a set T of triangles, (ii) if two different triangles in T have more than a vertex in common then they have exactly a side in common, and (iii) the set of vertices of the triangles in T is precisely the set of $(n+m)$ points $P_1, \cdots, Q_m$. How many triangles in T?

A-4. Show that

$$\int_0^1 x^x dx = \sum_{n=1}^{\infty} (-1)^{n+1} n^{-n}.$$

(The integrand is taken to be 1 at $x=0$.)

A-5. Let $u(t)$ be a continuous function in the system of differential equations

$$\frac{dx}{dt} = -2y + u(t), \qquad \frac{dy}{dt} = -2x + u(t).$$

Show that, regardless of the choice of $u(t)$, the solution of the system which satisfies $x=x_0$, $y=y_0$ at $t=0$ will never pass through (0, 0) unless $x_0=y_0$. When $x_0=y_0$, show that, for any positive value t_0 of t, it is possible to choose $u(t)$ so the solution is at (0, 0) when $t=t_0$.

A-6. Let a sequence $\{x_n\}$ be given, and let $y_n=x_{n-1}+2x_n$, $n=2, 3, 4, \cdots$. Suppose that the sequence $\{y_n\}$ converges. Prove that the sequence $\{x_n\}$ also converges.

B-1. Let n be a positive integer such that $n+1$ is divisible by 24. Prove that the sum of all the divisors of n is divisible by 24.

B-2. Show that a finite group can not be the union of two of its proper subgroups. Does the statement remain true if "two" is replaced by "three"?

B-3. The terms of a sequence T_n satisfy

$$T_n T_{n+1} = n \qquad (n = 1, 2, 3, \cdots) \text{ and } \lim_{n\to\infty} \frac{T_n}{T_{n+1}} = 1.$$

Show that $\pi T_1^2 = 2$.

B-4. Show that any curve of unit length can be covered by a closed rectangle of area 1/4.

B-5. Let $a_1 < a_2 < a_3 < \cdots$ be an increasing sequence of positive integers. Let the series

$$\sum_{n=1}^{\infty} 1/a_n$$

be convergent. For any number x, let $k(x)$ be the number of the a_n's which do not exceed x. Show that $\lim_{x\to\infty} k(x)/x = 0$.

B-6. Let A and B be matrices of size 3×2 and 2×3 respectively. Suppose that their product in the order AB is given by

$$AB = \begin{bmatrix} 8 & 2 & -2 \\ 2 & 5 & 4 \\ -2 & 4 & 5 \end{bmatrix}.$$

Show that the product BA is given by

$$BA = \begin{bmatrix} 9 & 0 \\ 0 & 9 \end{bmatrix}.$$

THE THIRTY-FIRST WILLIAM LOWELL PUTNAM MATHEMATICAL COMPETITION

December 5, 1970

A-1. Show that the power series for the function

$$e^{ax} \cos bx \qquad (a > 0, b > 0)$$

in powers of x has either no zero coefficients or infinitely many zero coefficients.

A-2. Consider the locus given by the real polynomial equation

$$Ax^2 + Bxy + Cy^2 + Dx^3 + Ex^2y + Fxy^2 + Gy^3 = 0,$$

where $B^2 - 4AC < 0$. Prove that there is a positive number δ such that there are no points of the locus in the punctured disk

$$0 < x^2 + y^2 < \delta^2.$$

A-3. Find the length of the longest sequence of equal nonzero digits in which an integral square can terminate (in base 10) and find the smallest square which terminates in such a sequence.

A-4. Given a sequence $\{x_n\}$, $n = 1, 2, \cdots$, such that $\text{limit}_{n\to\infty} \{x_n - x_{n-2}\} = 0$. Prove that

$$\underset{n\to\infty}{\text{limit}} \frac{x_n - x_{n-1}}{n} = 0.$$

A-5. Determine the radius of the largest circle which can lie on the ellipsoid

$$\frac{x^2}{a^2} + \frac{y^2}{b^2} + \frac{z^2}{c^2} = 1 \qquad (a > b > c).$$

A-6. Three numbers are chosen independently at random, one from each of the three intervals $[0, L_i]$ $(i = 1, 2, 3)$. If the distribution of each random number is uniform with respect to length in the interval it is chosen from, determine the expected value of the smallest of the three numbers chosen.

B-1. Evaluate

$$\underset{n\to\infty}{\text{limit}} \frac{1}{n^4} \prod_{i=1}^{2n} (n^2 + i^2)^{1/n}.$$

B-2. The time-varying temperature of a certain body is given by a polynomial in the time of degree at most three. Show that the average temperature of the body between 9 A.M. and 3 P.M. can always be found by taking the average of the temperatures at two fixed times, which are independent of which polynomial occurs. Also, show that these two times are 10:16 A.M. and 1:44 P.M. to the nearest minute.

B-3. A closed subset S of R^2 lies in $a < x < b$. Show that its projection on the y-axis is closed.

B-4. An automobile starts from rest and ends at rest, traversing a distance of one mile in one minute, along a straight road. If a governor prevents the speed of the car from exceeding ninety miles per hour, show that at some time of the traverse the acceleration or deceleration of the car was at least 6.6 ft./sec.^{2}

B-5. Let u_n denote the "ramp" function

$$u_n(x) = \begin{cases} -n & \text{for } x \leqq -n, \\ x & \text{for } -n < x \leqq n, \\ n & \text{for } x > n, \end{cases}$$

and let F denote a real function of a real variable. Show that F is continuous if and only if $u_n \circ F$ is continuous for all n. (Note: $(u_n \circ F)(x) = u_n[F(x)]$.)

B-6. A quadrilateral which can be inscribed in a circle is said to be *inscribable* or *cyclic*. A quadrilateral which can be circumscribed to a circle is said to be *circumscribable*. Show that if a circumscribable quadrilateral of sides a, b, c, d has area $A = \sqrt{abcd}$, then it is also inscribable.

THE THIRTY-SECOND WILLIAM LOWELL PUTNAM MATHEMATICAL COMPETITION

December 4, 1971

A–1. Let there be given nine lattice points (points with integral coordinates) in three dimensional Euclidean space. Show that there is a lattice point on the interior of one of the line segments joining two of these points.

A–2. Determine all polynomials $P(x)$ such that $P(x^2 + 1) = (P(x))^2 + 1$ and $P(0) = 0$.

A–3. The three vertices of a triangle of sides a, b, and c are lattice points and lie on a circle of radius R. Show that $abc \geqq 2R$. (Lattice points are points in the Euclidean plane with integral coordinates.)

A–4. Show that for $0 < \varepsilon < 1$ the expression $(x + y)^n (x^2 - (2-\varepsilon)xy + y^2)$ is a polynomial with positive coefficients for n sufficiently large and integral. For $\varepsilon = .002$ find the smallest admissible value of n.

A–5. A game of solitaire is played as follows. After each play, according to the outcome, the player receives either a or b points (a and b are positive integers with a greater than b), and his score accumulates from play to play. It has been noticed that there are thirty-five non-attainable scores and that one of these is 58. Find a and b.

A–6. Let c be a real number such that n^c is an integer for every positive integer n. Show that c is a non-negative integer.

B–1. Let S be a set and let $\circ$ be a binary operation on S satisfying the two laws

$$x \circ x = x \text{ for all } x \text{ in } S, \text{ and}$$

$$(x \circ y) \circ z = (y \circ z) \circ x \text{ for all } x, y, z \text{ in } S.$$

Show that $\circ$ is associative and commutative.

B–2. Let $F(x)$ be a real valued function defined for all real x except for $x = 0$ and $x = 1$ and satisfying the functional equation $F(x) + F\{(x-1)/x\} = 1 + x$. Find all functions $F(x)$ satisfying these conditions.

B–3. Two cars travel around a track at equal and constant speeds, each completing a lap every hour. From a common starting point, the first starts at time $t = 0$ and the second at an arbitrary later time $t = T > 0$. Prove that there is a total period of exactly one hour during the motion in which the first has completed twice as many laps as the second.

B–4. A "spherical ellipse" with foci A, B on a given sphere is defined as the set of all points P on the sphere such that $\overset{\frown}{PA} + \overset{\frown}{PB} = \text{constant}$. Here $\overset{\frown}{PA}$ denotes the shortest distance on the sphere between P and A. Determine the entire class of real spherical ellipses which are circles.

B–5. Show that the graphs in the x-y plane of all solutions of the system of differential equations

$$x'' + y' + 6x = 0, \; y'' - x' + 6y = 0 \quad (' = d/dt)$$

which satisfy $x'(0) = y'(0) = 0$ are hypocycloids, and find the radius of the fixed circle and the two possible values of the radius of the rolling circle for each such solution. (A hypocycloid is the path described by a fixed point on the circumference of a circle which rolls on the inside of a given fixed circle.)

B–6. Let $\delta(x)$ be the greatest odd divisor of the positive integer x. Show that $\left| \sum_{n=1}^{x} \delta(n)/n - 2x/3 \right| < 1$, for all positive integers x.

THE THIRTY-THIRD WILLIAM LOWELL PUTNAM MATHEMATICAL COMPETITION

December 2, 1972

A–1. Show that there are no four consecutive binomial coefficients $\binom{n}{r}, \binom{n}{r+1}, \binom{n}{r+2}, \binom{n}{r+3}$ (n, r integers > 0 and $r + 3 \leqq n$) which are in arithmetic progression.

A–2. Let S be a set and let $*$ be a binary operation on S satisfying the laws

$$x * (x * y) = y \quad \text{for all } x, y \text{ in } S,$$

$$(y * x) * x = y \text{ for all } x, y \text{ in } S.$$

Show that $*$ is commutative but not necessarily associative.

A–3. If for a sequence $x_1, x_2, x_3, \cdots, \lim_{n\to\infty} (x_1+x_2+\cdots+x_n)/n$ exists, call this limit the C-limit of the sequence. A function $f(x)$ from $[0, 1]$ to the reals is called a supercontinuous function on the interval $[0, 1]$ if the C-limit exists for the sequence $f(x_1), f(x_2), f(x_3), \cdots$ whenever the C-limit exists for the sequence $x_1, x_2, x_3 \cdots$. Find all supercontinuous functions on $[0, 1]$.

A–4. Of all ellipses inscribed in a square, show that the circle has the maximum perimeter.

A–5. Show that if n is an integer greater than 1, then n does not divide $2^n - 1$.

A–6. Let $f(x)$ be an integrable function in $0 \leqq x \leqq 1$ and suppose $\int_0^1 f(x)dx = 0, \int_0^1 x f(x)dx = 0, \cdots, \int_0^1 x^{n-1} f(x)dx = 0$ and $\int_0^1 x^n f(x)dx = 1$. Show that $|f(x)| \geqq 2^n(n+1)$ in a set of positive measure.

B–1. Show that the power series representation for the series $\sum_{n=0}^{\infty} (x^n(x-1)^{2n})/n!$ cannot have three consecutive zero coefficients.

B–2. A particle moving on a straight line starts from rest and attains a velocity v_0 after traversing a distance s_0. If the motion is such that the acceleration was never increasing, find the maximum time for the traverse.

B–3. Let A and B be two elements in a group such that $ABA = BA^2B$, $A^3 = 1$ and $B^{2n-1} = 1$ for some positive integer n. Prove $B = 1$.

B–4. Let n be an integer greater than 1. Show that there exists a polynomial $P(x, y, z)$ with integral coefficients such that $x \equiv P(x^n, x^{n+1}, x + x^{n+2})$.

B–5. If the opposite angles of a skew (non-planar) quadrilateral are equal in pairs, prove that the opposite sides are equal in pairs.

B–6. Let $n_1 < n_2 < n_3 < \cdots < n_k$ be a set of positive integers. Prove that the polynomial $1 + z^{n_1} + z^{n_2} + \cdots + z^{n_k}$ has no roots inside the circle $|z| < (\sqrt{5} - 1)/2$.

THE THIRTY-FOURTH WILLIAM LOWELL PUTNAM MATHEMATICAL COMPETITION

December 1, 1973

A–1. (a) Let ABC be any triangle. Let X, Y, Z be points on the sides BC, CA, AB respectively. Suppose the distances $\overline{BX} \leqq \overline{XC}$, $\overline{CY} \leqq \overline{YA}$, $\overline{AZ} \leqq \overline{ZB}$ (see Figure 1). Show that the area of the triangle XYZ is $\geqq (1/4)$ (area of triangle ABC).

(b) Let ABC be any triangle, and let X, Y, Z be points on the sides BC, CA, AB respectively (but without any assumption about the ratios of the distances $\overline{BX}/\overline{XC}$, etc.; see Figures 1 and 2). Using (a) or by any other method, show: One of the three corner triangles AZY, BXZ, CYX has an area $\leqq$ area of triangle XYZ.

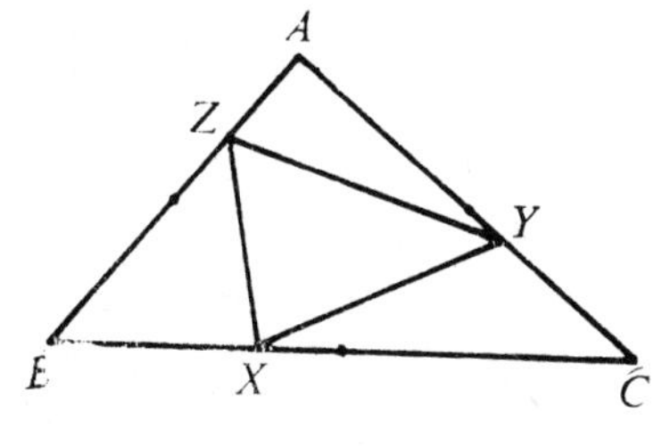

FIG. 1

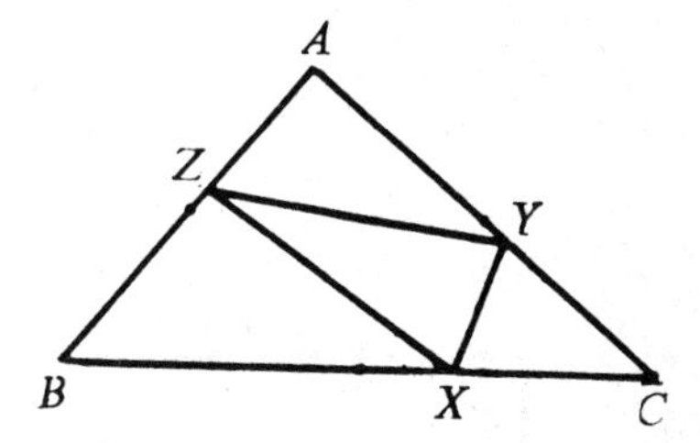

FIG. 2

A–2. Consider an infinite series whose nth term is $\pm (1/n)$, the $\pm$ signs being determined according to a pattern that repeats periodically in blocks of eight. [There are 2^8 possible patterns of which two examples are:

$$+ + - - - - + +,$$

$$+ - - - + - - - .$$

The first example would generate the series

$$1 + (1/2) - (1/3) - (1/4) - (1/5) - (1/6) + (1/7) + (1/8)$$

$$+ (1/9) + (1/10) - (1/11) - (1/12) - \cdots .]$$

(a) Show that a sufficient condition for the series to be conditionally convergent is that there be four "+" signs and four "−" signs in the block of eight.

(b) Is this sufficient condition also necessary?
[Here "convergent" means "convergent to a finite limit."]

A–3. Let n be a fixed positive integer and let $b(n)$ be the minimum value of

$$k + \frac{n}{k}$$

as k is allowed to range through all positive integers. Prove that $b(n)$ and $\sqrt{4n+1}$ have the same integer part. [The "integer part" of a real number is the greatest integer which does not exceed it, e.g. for π it is 3, for $\sqrt{21}$ it is 4, for 5 it is 5, etc.]

A–4. How many zeros does the function $f(x) = 2^x - 1 - x^2$ have on the real line? [By a "zero" of a function f, we mean a value x_0 in the domain of f (here the set of all real numbers) such that $f(x_0) = 0$.]

A–5. A particle moves in 3-space according to the equations:

$$\frac{dx}{dt} = yz, \quad \frac{dy}{dt} = zx, \quad \frac{dz}{dt} = xy.$$

[Here $x(t)$, $y(t)$, $z(t)$ are real-valued functions of the real variable t.] Show that:

(a) If two of $x(0)$, $y(0)$, $z(0)$ equal zero, then the particle never moves.

(b) If $x(0) = y(0) = 1$, $z(0) = 0$, then the solution is:

$$x = \sec t, \ y = \sec t, \ z = \tan t;$$

whereas if $x(0) = y(0) = 1$, $z(0) = -1$, then

$$x = 1/(t+1), y = 1/(t+1), z = -1/(t+1).$$

(c) If at least two of the values $x(0)$, $y(0)$, $z(0)$ are different from zero, then either the particle moves to infinity at some finite time in the future, or it came from infinity at some finite time in the past. [A point (x, y, z) in 3-space "moves to infinity" if its distance from the origin approaches infinity.]

A–6. Prove that it is impossible for seven distinct straight lines to be situated in the euclidean plane so as to have at least six points where exactly three of these lines intersect and at least four points where exactly two of these lines intersect.

B–1. Let $a_1, a_2, \ldots, a_{2n+1}$ be a set of integers such that, if any one of them is removed, the remaining ones can be divided into two sets of n integers with equal sums. Prove $a_1 = a_2 = \cdots = a_{2n+1}$.

B–2. Let $z = x + iy$ be a complex number with x and y rational and with $|z| = 1$. Show that the number $|z^{2n} - 1|$ is rational for every integer n.

B–3. Consider an integer $p > 1$ with the property that the polynomial $x^2 - x + p$ takes prime values for all integers x in the range $0 \leqq x < p$. (Examples: $p = 5$ and $p = 41$ have this property.) Show that there is exactly one triple of integers a, b, c satisfying the conditions:

$$b^2 - 4ac = 1 - 4p,$$

$$0 < a \leqq c,$$

$$-a \leqq b < a.$$

B–4. (a) On $[0, 1]$, let f have a continuous derivative satisfying $0 < f'(x) \leqq 1$. Also suppose that $f(0) = 0$. Prove that

$$\left[\int_0^1 f(x)dx\right]^2 \geqq \int_0^1 [f(x)]^3 dx.$$

[Hint: Replace the inequality by one involving the inverse function to f.]

(b) Show an example in which equality occurs.

B–5. (a) Let z be a solution of the quadratic equation

$$az^2 + bz + c = 0$$

and let n be a positive integer. Show that z can be expressed as a rational function of z^n, a, b, c.

(b) Using (a) or by any other means, express x as a rational function of x^3 and $x + (1/x)$. (Display your answer explicitly in a clearly visible form.)
[By a rational function of several variables, we mean a quotient of polynomials in those variables, the polynomials having rational numbers as coefficients, and the denominator being not identically zero. Thus to obtain x as a rational function of $u = x^2$ and $v = x + (1/x)$, we could write $x = (u + 1)/v$.]

B–6. On the domain $0 \leqq \theta \leqq 2\pi$:

(a) Prove that $\sin^2\theta \cdot \sin(2\theta)$ takes its maximum at $\pi/3$ and $4\pi/3$ (and hence its minimum at $2\pi/3$ and $5\pi/3$).

(b) Show that

$$\left| \sin^2\theta \{\sin^3(2\theta) \cdot \sin^3(4\theta) \cdots \sin^3(2^{n-1}\theta)\} \sin(2^n\theta) \right|$$

takes its maximum at $\theta = \pi/3$. (The maximum may also be attained at other points.)

(c) Derive the inequality:

$$\sin^2\theta \cdot \sin^2(2\theta) \cdot \sin^2(4\theta) \cdots \sin^2(2^n\theta) \leqq (3/4)^n.$$

THE THIRTY-FIFTH WILLIAM LOWELL PUTNAM MATHEMATICAL COMPETITION

December 7, 1974

A–1. Call a set of positive integers "conspiratorial" if no three of them are pairwise relatively prime. (A set of integers is "pairwise relatively prime" if no pair of them has a common divisor greater than 1.) What is the largest number of elements in any "conspiratorial" subset of the integers 1 through 16?

A–2. A circle stands in a plane perpendicular to the ground and a point A lies in this plane exterior to the circle and higher than its bottom. A particle starting from rest at A slides without friction down an inclined straight line until it reaches the circle. Which straight line allows descent in the shortest time? [Assume that the force of gravity is constant over the region involved, there are no relativistic effects, etc.]

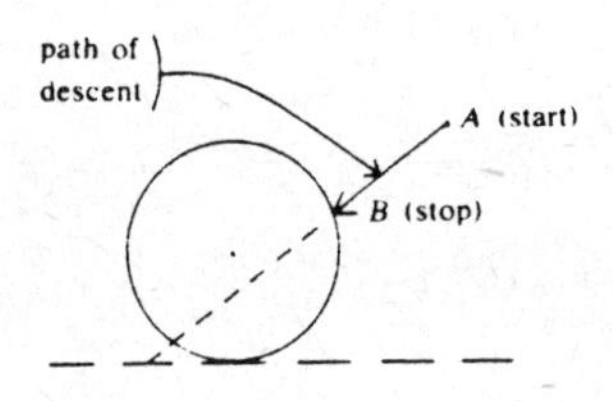

The starting point A and the circle are fixed; the stopping point B is allowed to vary over the circle.

Note. The answer may be given in any form which specifies the line of descent in an unambiguous manner; it is not required to find the coordinates of the point B.

A–3. A well-known theorem asserts that a prime $p > 2$ can be written as the sum of two perfect squares ($p = m^2 + n^2$, with m and n integers) if and only if $p \equiv 1 \pmod 4$. *Assuming* this result, find which primes $p > 2$ can be written in each of the following forms, using (not necessarily positive) integers x and y:
(a) $x^2 + 16y^2$;
(b) $4x^2 + 4xy + 5y^2$.

A–4. An unbiased coin is tossed n times. What is the expected value of $|H - T|$, where H is the number of heads and T is the number of tails? In other words, evaluate in *closed form*:

$$\frac{1}{2^{n-1}} \sum_{k<n/2} (n-2k)\binom{n}{k}.$$

(In this problem, "closed form" means a form not involving a series. The given series can be reduced to a single term involving only binomial coefficients, rational functions of n and 2^n, and the greatest integer function $[x]$.)

A–5. Consider the two mutually tangent parabolas $y = x^2$ and $y = -x^2$. [These have foci at $(0, 1/4)$ and $(0, -1/4)$, and directrices $y = -1/4$ and $y = 1/4$, respectively.] The upper parabola rolls without slipping around the fixed lower parabola. Find the locus of the focus of the moving parabola.

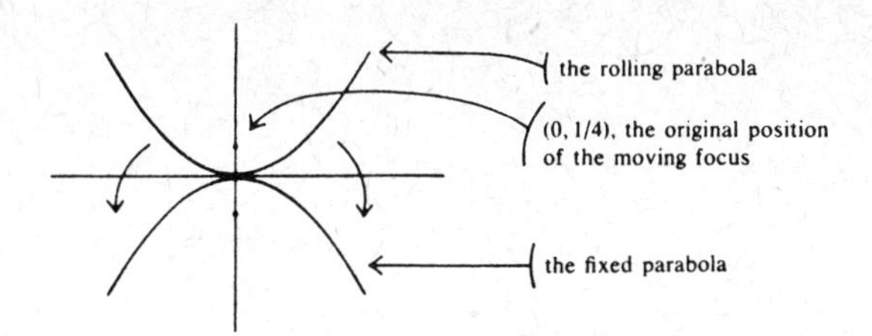

A–6. It is well known that the value of the polynomial $(x+1)(x+2)\cdots(x+n)$ is exactly divisible by n for every integer x. Given n, let $k = k(n)$ be the *minimal degree* of any monic integral polynomial

$$f(x) = x^k + a_1x^{k-1} + \cdots + a_k$$

(with integer coefficients and leading coefficient 1) such that the value of $f(x)$ is exactly divisible by n for every integer x.

Find the relationship between n and $k = k(n)$. In particular, find the value of k corresponding to $n = 1\,000\,000$.

B–1. Which configurations of five (not necessarily distinct) points $p_1, \cdots, p_5$ on the circle $x^2 + y^2 = 1$ maximize the sum of the ten distances

$$\sum_{i<j} d(p_i, p_j)\,?$$

[Here $d(p, q)$ denotes the straight line distance between p and q.]

B–2. Let $y(x)$ be a continuously differentiable real-valued function of a real variable x. Show that if $(y')^2 + y^3 \to 0$ as $x \to +\infty$, then $y(x)$ and $y'(x) \to 0$ as $x \to +\infty$.

B–3. Prove that if α is a real number such that

$$\cos \pi\alpha = 1/3,$$

then α is irrational. (The angle $\pi\alpha$ is in radians.)

B–4. In the standard definition, a real-valued function of two real variables $g: R^2 \to R^1$ is *continuous* if, for every point $(x_0, y_0) \in R^2$ and every $\varepsilon > 0$, there is a corresponding $\delta > 0$ such that $[(x - x_0)^2 + (y - y_0)^2]^{1/2} < \delta$ implies $|g(x, y) - g(x_0, y_0)| < \varepsilon$.

By contrast, $f: R^2 \to R^1$ is said to be *continuous in each variable separately* if, for each fixed value y_0 of y, the function $f(x, y_0)$ is continuous in the usual sense as a function of x, and similarly $f(x_0, y)$ is continuous as a function of y for each fixed x_0.

Let $f: R^2 \to R^1$ be continuous in each variable separately. Show that there exists a sequence of continuous functions $g_n: R^2 \to R^1$ such that

$$f(x, y) = \lim_{n\to\infty} g_n(x, y) \text{ for all } (x, y) \in R^2.$$

B–5. Show that $1 + (n/1!) + (n^2/2!) + \cdots + (n^n/n!) > e^n/2$ for every integer $n \geqq 0$.

REMARKS. You may assume as known Taylor's remainder formula:

$$e^x - \sum_{k=0}^{n} \frac{x^k}{k!} = \frac{1}{n!}\int_0^x (x - t)^n e^t\, dt,$$

as well as the fact that

$$n! = \int_0^\infty t^n e^{-t}\, dt.$$

B–6. For a set with n elements, how many subsets are there whose cardinality (the number of elements in the subset) is respectively $\equiv 0 \pmod 3$, $\equiv 1 \pmod 3$, $\equiv 2 \pmod 3$? In other words, calculate

$$s_{i,n} = \sum_{k \equiv i \pmod 3} \binom{n}{k} \qquad \text{for } i = 0, 1, 2.$$

Your result should be strong enough to permit direct evaluation of the numbers $s_{i,n}$ and to show clearly the relationship of $s_{0,n}$ and $s_{1,n}$ and $s_{2,n}$ to each other for all positive integers n. In particular, show the relationships among these three sums for $n = 1000$. [An illustration of the definition of $s_{i,n}$ is $s_{0,6} = \binom{6}{0} + \binom{6}{3} + \binom{6}{6} = 22$.]

THE THIRTY-SIXTH WILLIAM LOWELL PUTNAM MATHEMATICAL COMPETITION

December 6, 1975

A–1. Supposing that an integer n is the sum of two triangular numbers,

$$n = \frac{a^2 + a}{2} + \frac{b^2 + b}{2},$$

write $4n + 1$ as the sum of two squares, $4n + 1 = x^2 + y^2$, and show how x and y can be expressed in terms of a and b.

Show that, conversely, if $4n + 1 = x^2 + y^2$, then n is the sum of two triangular numbers.

[Of course, a, b, x, y are understood to be integers.]

A–2. For which ordered pairs of *real* numbers b, c do both roots of the quadratic equation

$$z^2 + bz + c = 0$$

lie inside the unit disk $\{|z| < 1\}$ in the complex plane?

Draw a reasonably accurate picture (i.e., 'graph') of the region in the real bc-plane for which the above condition holds. Identify precisely the boundary curves of this region.

A–3. Let a, b, c be constants with $0 < a < b < c$. At what points of the set

$$\{x^b + y^b + z^b = 1,\ x \geqq 0,\ y \geqq 0,\ z \geqq 0\}$$

in three-dimensional space R^3 does the function $f(x, y, z) = x^a + y^b + z^c$ assume its maximum and minimum values?

A–4. Let $n = 2m$, where m is an odd integer greater than 1. Let $\theta = e^{2\pi i/n}$. Express $(1 - \theta)^{-1}$ explicitly as a polynomial in θ,

$$a_k\theta^k + a_{k-1}\theta^{k-1} + \cdots + a_1\theta + a_0,$$

with *integer* coefficients a_i.

[Note that θ is a primitive n-th root of unity, and thus it satisfies all of the identities which hold for such roots.]

A–5. On some interval I of the real line, let $y_1(x)$ and $y_2(x)$ be linearly independent solutions of the differential equation

$$y'' = f(x)y,$$

where $f(x)$ is a continuous real-valued function. Suppose that $y_1(x) > 0$ and $y_2(x) > 0$ on I. Show that there exists a positive constant c such that, on I, the function

$$z(x) = c\sqrt{y_1(x)y_2(x)}$$

satisfies the equation

$$z'' + \frac{1}{z^3} = f(x)z.$$

State clearly the manner in which c depends on $y_1(x)$ and $y_2(x)$.

A–6. Let P_1, P_2, P_3 be the vertices of an acute-angled triangle situated in three-dimensional space. Show that it is always possible to locate two additional points P_4 and P_5 in such a way that no three of the points are collinear and so that the line through any two of the five points is perpendicular to the plane determined by the other three.

In writing your answer, state clearly the locations at which you place the points P_4 and P_5.

B–1. In the additive group of ordered pairs of integers (m, n) [with addition defined componentwise: $(m, n)+(m', n')=(m+m', n+n')$] consider the subgroup H generated by the three elements

$$(3,8), \qquad (4,-1), \qquad (5,4).$$

Then H has another set of generators of the form

$$(1, b), \qquad (0, a)$$

for some integers a, b with $a>0$. Find a.

[Elements $g_1, \ldots, g_k$ are said to *generate* a subgroup H if (i) each $g_i \in H$, and (ii) every $h \in H$ can be written as a sum $h = n_1 g_1 + \cdots + n_k g_k$ where the n_i are integers (and where, for example, $3g_1 - 2g_2$ means $g_1 + g_1 + g_1 - g_2 - g_2$).]

B–2. In three-dimensional Euclidean space, define a *slab* to be the open set of points lying between two parallel planes. The distance between the planes is called the *thickness* of the slab. Given an infinite sequence $S_1, S_2, \ldots$ of slabs of thicknesses $d_1, d_2, \ldots$, respectively, such that $\Sigma_{i=1}^{\infty} d_i$ converges, prove that there is some point in the space which is not contained in any of the slabs.

B–3. Let $s_k(a_1, \ldots, a_n)$ denote the k-th elementary symmetric function of $a_1, \ldots, a_n$. With k held fixed, find the supremum (or least upper bound) M_k of

$$s_k(a_1, \ldots, a_n)/[s_1(a_1, \ldots, a_n)]^k$$

for arbitrary $n \geqq k$ and arbitrary n-tuples $a_1, \ldots, a_n$ of positive real numbers.

[The symmetric function $s_k(a_1, \ldots, a_n)$ is the sum of all k-fold products of the variables $a_1, \ldots, a_n$. Thus, for example:

$$s_1(a_1, \ldots, a_n) = a_1 + a_2 + \ldots + a_n;$$

$$s_3(a_1, a_2, a_3, a_4) = a_1a_2a_3 + a_1a_2a_4 + a_1a_3a_4 + a_2a_3a_4.$$

It should be remarked that the supremum M_k is never attained; it is approached arbitrarily closely when, for fixed k, the number n of variables increases without bound, and the values $a_i > 0$ are suitably chosen.]

B–4. Does there exist a subset B of the unit circle $x^2 + y^2 = 1$ such that (i) B is topologically closed, and (ii) B contains exactly one point from each pair of diametrically opposite points on the circle?

[A set B is topologically *closed* if it contains the limit of every convergent sequence of points in B.]

B–5. Let $f_0(x) = e^x$ and $f_{n+1}(x) = x f_n'(x)$ for $n = 0, 1, 2, \ldots$. Show that

$$\sum_{n=0}^{\infty} \frac{f_n(1)}{n!} = e^e.$$

B–6. Show that if $s_n = 1 + \frac{1}{2} + \frac{1}{3} + \cdots + 1/n$, then

(a) $n(n+1)^{1/n} < n + s_n$ for $n > 1$, and

(b) $(n-1)n^{-1/(n-1)} < n - s_n$ for $n > 2$.

THE THIRTY-SEVENTH WILLIAM LOWELL PUTNAM MATHEMATICAL COMPETITION

December 4, 1976

A–1. P is an interior point of the angle whose sides are the rays **OA** and **OB**. Locate X on **OA** and Y on **OB** so that the line segment $\overline{XY}$ contains P and so that the product of distances $(PX)(PY)$ is a minimum.

A–2. Let $P(x, y) = x^2y + xy^2$ and $Q(x, y) = x^2 + xy + y^2$. For $n = 1, 2, 3, \ldots$, let $F_n(x, y) = (x + y)^n - x^n - y^n$ and $G_n(x, y) = (x + y)^n + x^n + y^n$. One observes that $G_2 = 2Q$, $F_3 = 3P$, $G_4 = 2Q^2$, $F_5 = 5PQ$, $G_6 = 2Q^3 + 3P^2$. Prove that, in fact, for each n either F_n or G_n is expressible as a polynomial in P and Q with integer coefficients.

A–3. Find all integral solutions of the equation

$$|p^r - q^s| = 1,$$

where p and q are prime numbers and r and s are positive integers larger than unity. Prove that there are no other solutions.

A–4. Let r be a root of $P(x) = x^3 + ax^2 + bx - 1 = 0$ and $r + 1$ be a root of $y^3 + cy^2 + dy + 1 = 0$, where a, b, c, and d are integers. Also let $P(x)$ be irreducible over the rational numbers. Express another root s of $P(x) = 0$ as a function of r which does not explicitly involve a, b, c, or d.

A–5. In the (x, y)-plane, if R is the set of points inside and on a convex polygon, let $D(x, y)$ be the distance from (x, y) to the nearest point of R. (a) Show that there exist constants a, b, and c, independent of R, such that

$$\int_{-\infty}^{\infty}\int_{-\infty}^{\infty} e^{-D(x,y)}dx\,dy = a + bL + cA,$$

where L is the perimeter of R and A is the area of R. (b) Find the values of a, b, and c.

A–6. Suppose $f(x)$ is a twice continuously differentiable real valued function defined for all real numbers x and satisfying $|f(x)| \leqq 1$ for all x and $(f(0))^2 + (f'(0))^2 = 4$. Prove that there exists a real number x_0 such that $f(x_0) + f''(x_0) = 0$.

B–1. Evaluate

$$\lim_{n\to\infty} \frac{1}{n} \sum_{k=1}^{n} \left(\left[\frac{2n}{k}\right] - 2\left[\frac{n}{k}\right]\right)$$

and express your answer in the form $\log a - b$, with a and b positive integers.

Here $[x]$ is defined to be the integer such that $[x] \leqq x < [x] + 1$ and $\log x$ is the logarithm of x to base e.

B–2. Suppose that G is a group generated by elements A and B, that is, every element of G can be written as a finite "word" $A^{n_1}B^{n_2}A^{n_3}\cdots B^{n_k}$, where $n_1, \ldots, n_k$ are any integers, and $A^0 = B^0 = 1$ as usual. Also, suppose that $A^4 = B^7 = ABA^{-1}B = 1$, $A^2 \neq 1$, and $B \neq 1$.

(a) How many elements of G are of the form C^2 with C in G?

(b) Write each such square as a word in A and B.

B–3. Suppose that we have n events $A_1, \ldots, A_n$, each of which has probability at least $1 - a$ of occurring, where $a < 1/4$. Further suppose that A_i and A_j are mutually independent if $|i - j| > 1$, although A_i and A_{i+1} may be dependent. Assume as known that the recurrence $u_{k+1} = u_k - au_{k-1}$, $u_0 = 1$, $u_1 = 1 - a$, defines positive real numbers u_k for $k = 0, 1, \ldots$. Show that the probability of all of $A_1, \ldots, A_n$ occurring is at least u_n.

B–4. For a point P on an ellipse, let d be the distance from the center of the ellipse to the line tangent to the ellipse at P. Prove that $(PF_1)(PF_2)d^2$ is constant as P varies on the ellipse, where PF_1 and PF_2 are the distances from P to the foci F_1 and F_2 of the ellipse.

B–5. Evaluate

$$\sum_{k=0}^{n} (-1)^k \binom{n}{k} (x-k)^n.$$

B–6. As usual, let $\sigma(N)$ denote the sum of all the (positive integral) divisors of N. (Included among these divisors are 1 and N itself.) For example, if p is a prime, then $\sigma(p) = p + 1$. Motivated by the notion of a "perfect" number, a positive integer N is called "quasiperfect" if $\sigma(N) = 2N + 1$. Prove that every quasiperfect number is the square of an odd integer.

THE THIRTY-EIGHTH WILLIAM LOWELL PUTNAM MATHEMATICAL COMPETITION

December 3, 1977

Problem A-1

Consider all lines which meet the graph of

$$y=2x^4+7x^3+3x-5$$

in four distinct points, say (x_i,y_i), $i=1,2,3,4$. Show that

$$\frac{x_1+x_2+x_3+x_4}{4}$$

is independent of the line and find its value.

Problem A-2

Determine all solutions in real numbers x,y,z,w of the system

$$x+y+z=w,$$

$$\frac{1}{x}+\frac{1}{y}+\frac{1}{z}=\frac{1}{w}.$$

Problem A-3

Let u,f, and g be functions, defined for all real numbers x, such that

$$\frac{u(x+1)+u(x-1)}{2}=f(x) \quad \text{and} \quad \frac{u(x+4)+u(x-4)}{2}=g(x).$$

Determine $u(x)$ in terms of f and g.

Problem A-4

For $0<x<1$, express

$$\sum_{n=0}^{\infty}\frac{x^{2^n}}{1-x^{2^{n+1}}}$$

as a rational function of x.

Problem A-5

Prove that

$$\binom{pa}{pb}\equiv\binom{a}{b}(\bmod p)$$

for all integers p,a, and b with p a prime, $p>0$, and $a\geqslant b\geqslant 0$.

Notation: $\binom{m}{n}$ denotes the binomial coefficient $\dfrac{m!}{n!(m-n)!}$.

Problem A-6

Let $f(x,y)$ be a continuous function on the square

$$S=\{(x,y):0\leqslant x\leqslant 1,0\leqslant y\leqslant 1\}.$$

For each point (a,b) in the interior of S, let $S_{(a,b)}$ be the largest square that is contained in S, is centered at (a,b), and has sides parallel to those of S. If the double integral $\iint f(x,y)\,dx\,dy$ is zero when taken over each square

$S_{(a,b)}$, must $f(x,y)$ be identically zero on S?

Problem B-1

Evaluate the infinite product

$$\prod_{n=2}^{\infty} \frac{n^3-1}{n^3+1}.$$

Problem B-2

Given a convex quadrilateral $ABCD$ and a point O not in the plane of $ABCD$, locate point A' on line OA, point B' on line OB, point C' on line OC, and point D' on line OD so that $A'B'C'D'$ is a parallelogram.

Problem B-3

An (ordered) triple (x_1,x_2,x_3) of positive *irrational* numbers with $x_1+x_2+x_3=1$ is called "balanced" if each $x_i<1/2$. If a triple is not balanced, say if $x_j>1/2$, one performs the following "balancing act"

$$B(x_1,x_2,x_3)=(x_1',x_2',x_3'),$$

where $x_i'=2x_i$ if $i\neq j$ and $x_j'=2x_j-1$. If the new triple is not balanced, one performs the balancing act on it. Does continuation of this process always lead to a balanced triple after a finite number of performances of the balancing act?

Problem B-4

Let C be a continuous closed curve in the plane which does not cross itself and let Q be a point inside C. Show that there exist points P_1 and P_2 on C such that Q is the midpoint of the line segment P_1P_2.

Problem B-5

Suppose that $a_1, a_2, \ldots, a_n$ are real $(n>1)$ and

$$A+\sum_{i=1}^{n} a_i^2 < \frac{1}{n-1}\left(\sum_{i=1}^{n} a_i\right)^2.$$

Prove that $A<2a_ia_j$ for $1\leq i<j\leq n$.

Problem B-6

Let H be a subgroup with h elements in a group G. Suppose that G has an element a such that for all x in H, $(xa)^3=1$, the identity. In G, let P be the subset of all products $x_1ax_2a\cdots x_na$, with n a positive integer and the x_i in H.

(a) Show that P is a finite set.

(b) Show that, in fact, P has no more than $3h^2$ elements.

THE THIRTY-NINTH WILLIAM LOWELL PUTNAM MATHEMATICAL COMPETITION

December 2, 1978

Problem A-1

Let A be any set of 20 distinct integers chosen from the arithmetic progression $1, 4, 7, \ldots, 100$. Prove that there must be two distinct integers in A whose sum is 104.

Problem A-2

Let $a, b, p_1, p_2, \ldots, p_n$ be real numbers with $a \neq b$. Define $f(x) = (p_1 - x)(p_2 - x)(p_3 - x) \cdots (p_n - x)$. Show that

$$\det \begin{pmatrix} p_1 & a & a & a & \cdots & a & a \\ b & p_2 & a & a & \cdots & a & a \\ b & b & p_3 & a & \cdots & a & a \\ b & b & b & p_4 & \cdots & a & a \\ \vdots & \vdots & \vdots & \vdots & & \vdots & \vdots \\ b & b & b & b & \cdots & p_{n-1} & a \\ b & b & b & b & \cdots & b & p_n \end{pmatrix} = \frac{bf(a) - af(b)}{b - a}.$$

Problem A-3

Let $p(x) = 2 + 4x + 3x^2 + 5x^3 + 3x^4 + 4x^5 + 2x^6$. For k with $0 < k < 5$, define

$$I_k = \int_0^\infty \frac{x^k}{p(x)}\,dx.$$

For which k is I_k smallest?

Problem A-4

A "bypass" operation on a set S is a mapping from $S \times S$ to S with the property

$$B(B(w,x), B(y,z)) = B(w,z) \qquad \text{for all} \quad w, x, y, z \text{ in } S.$$

(a) Prove that $B(a,b) = c$ implies $B(c,c) = c$ when B is a bypass.
(b) Prove that $B(a,b) = c$ implies $B(a,x) = B(c,x)$ for all x in S when B is a bypass.
(c) Construct a table for a bypass operation B on a finite set S with the following three properties:
  (i) $B(x,x) = x$ for all x in S.
  (ii) There exist d and e in S with $B(d,e) = d \neq e$.
  (iii) There exist f and g in S with $B(f,g) \neq f$.

Problem A-5

Let $0 < x_i < \pi$ for $i = 1, 2, \ldots, n$ and set

$$x = \frac{x_1 + x_2 + \cdots + x_n}{n}.$$

Prove that

$$\prod_{i=1}^{n} \frac{\sin x_i}{x_i} \leq \left(\frac{\sin x}{x}\right)^n.$$

Problem A-6

Let n distinct points in the plane be given. Prove that fewer than $2n^{3/2}$ pairs of them are unit distance apart.

Problem B-1

Find the area of a convex octagon that is inscribed in a circle and has four consecutive sides of length 3 units and the remaining four sides of length 2 units. Give the answer in the form $r+s\sqrt{t}$ with r, s, and t positive integers.

Problem B-2

Express

$$\sum_{n=1}^{\infty}\sum_{m=1}^{\infty}\frac{1}{m^2n+mn^2+2mn}$$

as a rational number.

Problem B-3

The sequence $\{Q_n(x)\}$ of polynomials is defined by

$$Q_1(x)=1+x,\ Q_2(x)=1+2x,$$

and, for $m>1$, by

$$Q_{2m+1}(x)=Q_{2m}(x)+(m+1)xQ_{2m-1}(x),$$
$$Q_{2m+2}(x)=Q_{2m+1}(x)+(m+1)xQ_{2m}(x).$$

Let x_n be the largest real solution of $Q_n(x)=0$. Prove that $\{x_n\}$ is an increasing sequence and that $\lim_{n\to\infty}x_n=0$.

Problem B-4

Prove that for every real number N, the equation

$$x_1^2+x_2^2+x_3^2+x_4^2=x_1x_2x_3+x_1x_2x_4+x_1x_3x_4+x_2x_3x_4$$

has a solution for which x_1, x_2, x_3, x_4 are all integers larger than N.

Problem B-5

Find the largest A for which there exists a polynomial

$$P(x)=Ax^4+Bx^3+Cx^2+Dx+E,$$

with real coefficients, which satisfies

$$0\le P(x)\le 1 \quad \text{for} \quad -1\le x\le 1.$$

Problem B-6

Let p and n be positive integers. Suppose that the numbers $c_{h,k}$ $(h=1,2,\ldots,n;\ k=1,2,\ldots,ph)$ satisfy $0\le c_{h,k}\le 1$. Prove that

$$\left(\sum\frac{c_{h,k}}{h}\right)^2\le 2p\sum c_{h,k},$$

where each summation is over all admissible ordered pairs (h,k).

THE FORTIETH WILLIAM LOWELL PUTNAM MATHEMATICAL COMPETITION

December 1, 1979

Problem A-1

Find positive integers n and $a_1, a_2, \ldots, a_n$ such that

$$a_1 + a_2 + \cdots + a_n = 1979$$

and the product $a_1 a_2 \cdots a_n$ is as large as possible.

Problem A-2

Establish necessary and sufficient conditions on the constant k for the existence of a continuous real valued function $f(x)$ satisfying $f(f(x)) = kx^9$ for all real x.

Problem A-3

Let $x_1, x_2, x_3, \ldots$ be a sequence of nonzero real numbers satisfying

$$x_n = \frac{x_{n-2}x_{n-1}}{2x_{n-2} - x_{n-1}} \quad \text{for } n = 3, 4, 5, \ldots.$$

Establish necessary and sufficient conditions on x_1 and x_2 for x_n to be an integer for infinitely many values of n.

Problem A-4

Let A be a set of $2n$ points in the plane, no three of which are collinear. Suppose that n of them are colored red and the remaining n blue. Prove or disprove: there are n closed straight line segments, no two with a point in common, such that the endpoints of each segment are points of A having different colors.

Problem A-5

Denote by $[x]$ the greatest integer less than or equal to x and by $S(x)$ the sequence $[x], [2x], [3x], \ldots$. Prove that there are distinct real solutions α and β of the equation $x^3 - 10x^2 + 29x - 25 = 0$ such that infinitely many positive integers appear both in $S(\alpha)$ and in $S(\beta)$.

Problem A-6

Let $0 \le p_i \le 1$ for $i = 1, 2, \ldots, n$. Show that

$$\sum_{i=1}^{n} \frac{1}{|x - p_i|} \le 8n\left(1 + \frac{1}{3} + \frac{1}{5} + \cdots + \frac{1}{2n-1}\right)$$

for some x satisfying $0 \le x \le 1$.

Problem B-1

Prove or disprove: there is at least one straight line normal to the graph of $y = \cosh x$ at a point $(a, \cosh a)$ and also normal to the graph of $y = \sinh x$ at a point $(c, \sinh c)$.

[At a point on a graph, the normal line is the perpendicular to the tangent at that point. Also, $\cosh x = (e^x + e^{-x})/2$ and $\sinh x = (e^x - e^{-x})/2$.]

Problem B-2

Let $0 < a < b$. Evaluate

$$\lim_{t \to 0} \left\{ \int_0^1 [bx + a(1-x)]^t \, dx \right\}^{1/t}$$

[The final answer should not involve any operations other than addition, subtraction, multiplication, division, and exponentiation.]

Problem B-3

Let F be a finite field having an odd number m of elements. Let $p(x)$ be an irreducible (i.e., nonfactorable) polynomial over F of the form

$$x^2+bx+c, \qquad b,c\in F.$$

For how many elements k in F is $p(x)+k$ irreducible over F?

Problem B-4

(a) Find a solution that is not identically zero, of the homogeneous linear differential equation

$$(3x^2+x-1)y''-(9x^2+9x-2)y'+(18x+3)y=0.$$

Intelligent guessing of the form of a solution may be helpful.

(b) Let $y=f(x)$ be the solution of the *nonhomogeneous* differential equation

$$(3x^2+x-1)y''-(9x^2+9x-2)y'+(18x+3)y=6(6x+1)$$

that has $f(0)=1$ and $(f(-1)-2)(f(1)-6)=1$. Find integers a, b, c such that $(f(-2)-a)(f(2)-b)=c$.

Problem B-5

In the plane, let C be a closed convex set that contains $(0,0)$ but no other point with integer coordinates. Suppose that $A(C)$, the area of C, is equally distributed among the four quadrants. Prove that $A(C)\leq 4$.

Problem B-6

For $k=1,2,\ldots,n$ let $z_k=x_k+iy_k$, where the x_k and y_k are real and $i=\sqrt{-1}$. Let r be the absolute value of the *real part* of

$$\pm\sqrt{z_1^2+z_2^2+\cdots+z_n^2}\ .$$

Prove that $r\leq |x_1|+|x_2|+\cdots+|x_n|$.

THE FORTY-FIRST WILLIAM LOWELL PUTNAM MATHEMATICAL COMPETITION

December 6, 1980

Problem A-1

Let b and c be fixed real numbers and let the ten points (j, y_j), $j = 1, 2, \ldots, 10$, lie on the parabola $y = x^2 + bx + c$. For $j = 1, 2, \ldots, 9$, let I_j be the point of intersection of the tangents to the given parabola at (j, y_j) and $(j + 1, y_{j+1})$. Determine the polynomial function $y = g(x)$ of least degree whose graph passes through all nine points I_j.

Problem A-2

Let r and s be positive integers. Derive a formula for the number of ordered quadruples (a, b, c, d) of positive integers such that

$$3^r \cdot 7^s = \operatorname{lcm}[a, b, c] = \operatorname{lcm}[a, b, d] = \operatorname{lcm}[a, c, d] = \operatorname{lcm}[b, c, d].$$

The answer should be a function of r and s.

(Note that 1 cm $[x, y, z]$ denotes the least common multiple of x, y, z.)

Problem A-3

Evaluate

$$\int_0^{\pi/2} \frac{dx}{1 + (\tan x)^{\sqrt{2}}}.$$

Problem A-4

(a) Prove that there exist integers a, b, c, not all zero and each of absolute value less than one million, such that

$$|a + b\sqrt{2} + c\sqrt{3}| < 10^{-11}.$$

(b) Let a, b, c be integers, not all zero and each of absolute value less than one million. Prove that

$$|a + b\sqrt{2} + c\sqrt{3}| > 10^{-21}.$$

Problem A-5

Let $P(t)$ be a nonconstant polynomial with real coefficients. Prove that the system of simultaneous equations

$$0 = \int_0^x P(t)\sin t\, dt = \int_0^x P(t)\cos t\, dt$$

has only finitely many real solutions x.

Problem A-6

Let C be the class of all real valued continuously differentiable functions f on the interval $0 \leqslant x \leqslant 1$ with $f(0) = 0$ and $f(1) = 1$. Determine the largest real number u such that

$$u \leqslant \int_0^1 |f'(x) - f(x)|\, dx$$

for all f in C.

Problem B-1

For which real numbers c is $(e^x + e^{-x})/2 \leqslant e^{cx^2}$ for all real x?

Problem B-2

Let S be the solid in three-dimensional space consisting of all points (x, y, z) satisfying the following system of six simultaneous conditions:

$$x \geqslant 0, \quad y \geqslant 0, \quad z \geqslant 0,$$

$$x + y + z \leqslant 11,$$

$$2x + 4y + 3z \leqslant 36,$$

$$2x + 3z \leqslant 24.$$

(a) Determine the number v of vertices of S.

(b) Determine the number e of edges of S.

(c) Sketch in the bc-plane the set of points (b, c) such that $(2, 5, 4)$ is one of the points (x, y, z) at which the linear function $bx + cy + z$ assumes its maximum value on S.

Problem B-3

For which real numbers a does the sequence defined by the initial condition $u_0 = a$ and the recursion $u_{n+1} = 2u_n - n^2$ have $u_n > 0$ for all $n \geqslant 0$?

(Express the answer in the simplest form.)

Problem B-4

Let $A_1, A_2, \ldots, A_{1066}$ be subsets of a finite set X such that $|A_i| > \frac{1}{2}|X|$ for $1 \leqslant i \leqslant 1066$. Prove there exist ten elements $x_1, \ldots, x_{10}$ of X such that every A_i contains at least one of $x_1, \ldots, x_{10}$.

(Here $|S|$ means the number of elements in the set S.)

Problem B-5

For each $t \geqslant 0$, let S_t be the set of all nonnegative, increasing, convex, continuous, real-valued functions $f(x)$ defined on the closed interval $[0, 1]$ for which

$$f(1) - 2f(2/3) + f(1/3) \geqslant t[f(2/3) - 2f(1/3) + f(0)].$$

Develop necessary and sufficient conditions on t for S_t to be closed under multiplication.

(This closure means that, if the functions $f(x)$ and $g(x)$ are in S_t, so is their product $f(x)g(x)$. A function $f(x)$ is convex if and only if $f(su + (1 - s)v) \leqslant sf(u) + (1 - s)f(v)$ whenever $0 \leqslant s \leqslant 1$.)

Problem B-6

An infinite array of rational numbers $G(d, n)$ is defined for integers d and n with $1 \leqslant d \leqslant n$ as follows:

$$G(1, n) = \frac{1}{n}, \qquad G(d, n) = \frac{d}{n} \sum_{i=d}^{n} G(d-1, i-1) \quad \text{for} \quad d > 1.$$

For $1 < d \leqslant p$ and p *prime*, prove that $G(d, p)$ is expressible as a quotient s/t of integers s and t with t *not* an integral multiple of p.

(For example, $G(3, 5) = 7/4$ with the denominator 4 not a multiple of 5.)

THE FORTY-SECOND WILLIAM LOWELL PUTNAM MATHEMATICAL COMPETITION

December 5, 1981

Problem A-1

Let $E(n)$ denote the largest integer k such that 5^k is an integral divisor of the product $1^1 2^2 3^3 \cdots n^n$. Calculate

$$\lim_{n \to \infty} \frac{E(n)}{n^2}.$$

Problem A-2

Two distinct squares of the 8 by 8 chessboard C are said to be adjacent if they have a vertex or side in common. Also, g is called a C-gap if for every numbering of the squares of C with all the integers $1, 2, \ldots, 64$ there exist two adjacent squares whose numbers differ by at least g. Determine the largest C-gap g.

Problem A-3

Find

$$\lim_{t \to \infty} \left[e^{-t} \int_0^t \int_0^t \frac{e^x - e^y}{x - y} \, dx \, dy \right]$$

or show that the limit does not exist.

Problem A-4

A point P moves inside a unit square in a straight line at unit speed. When it meets a corner it escapes. When it meets an edge its line of motion is reflected so that the angle of incidence equals the angle of reflection.

Let $N(T)$ be the number of starting directions from a fixed interior point P_0 for which P escapes within T units of time. Find the least constant a for which constants b and c exist such that

$$N(T) \leq aT^2 + bT + c$$

for all $T > 0$ and all initial points P_0.

Problem A-5

Let $P(x)$ be a polynomial with real coefficients and form the polynomial

$$Q(x) = (x^2 + 1) P(x) P'(x) + x\left([P(x)]^2 + [P'(x)]^2\right).$$

Given that the equation $P(x) = 0$ has n distinct real roots exceeding 1, prove or disprove that the equation $Q(x) = 0$ has at least $2n - 1$ distinct real roots.

Problem A-6

Suppose that each of the vertices of ΔABC is a lattice point in the (x, y)-plane and that there is exactly one lattice point P in the *interior* of the triangle. The line AP is extended to meet BC at E. Determine the largest possible value for the ratio of lengths of segments

$$\frac{|AP|}{|PE|}.$$

[A lattice point is a point whose coordinates x and y are integers.]

Problem B-1

Find

$$\lim_{n\to\infty}\left[\frac{1}{n^5}\sum_{h=1}^{n}\sum_{k=1}^{n}(5h^4-18h^2k^2+5k^4)\right].$$

Problem B-2

Determine the minimum value of

$$(r-1)^2+\left(\frac{s}{r}-1\right)^2+\left(\frac{t}{s}-1\right)^2+\left(\frac{4}{t}-1\right)^2$$

for all real numbers r, s, t with $1 \le r \le s \le t \le 4$.

Problem B-3

Prove that there are infinitely many positive integers n with the property that if p is a prime divisor of n^2+3, then p is also a divisor of k^2+3 for some integer k with $k^2 < n$.

Problem B-4

Let V be a set of 5 by 7 matrices, with real entries and with the property that $rA + sB \in V$ whenever $A, B \in V$ and r and s are scalars (i.e., real numbers). *Prove or disprove* the following assertion: If V contains matrices of ranks 0, 1, 2, 4, and 5, then it also contains a matrix of rank 3.

[The rank of a nonzero matrix M is the largest k such that the entries of some k rows and some k columns form a k by k matrix with a nonzero determinant.]

Problem B-5

Let $B(n)$ be the number of ones in the base two expression for the positive integer n. For example, $B(6) = B(110_2) = 2$ and $B(15) = B(1111_2) = 4$. Determine whether or not

$$\exp\left(\sum_{n=1}^{\infty}\frac{B(n)}{n(n+1)}\right)$$

is a rational number. Here $\exp(x)$ denotes e^x.

Problem B-6

Let C be a fixed unit circle in the Cartesian plane. For any convex polygon P each of whose sides is tangent to C, let $N(P, h, k)$ be the number of points common to P and the unit circle with center at (h, k). Let $H(P)$ be the region of all points (x, y) for which $N(P, x, y) \ge 1$ and $F(P)$ be the area of $H(P)$. Find the smallest number u with

$$\frac{1}{F(P)}\iint N(P, x, y)\,dx\,dy < u$$

for all polygons P, where the double integral is taken over $H(P)$.

THE FORTY-THIRD WILLIAM LOWELL PUTNAM MATHEMATICAL COMPETITION

December 4, 1982

Problem A-1

Let V be the region in the cartesian plane consisting of all points (x, y) satisfying the simultaneous conditions

$$|x| \leqslant y \leqslant |x| + 3 \quad \text{and} \quad y \leqslant 4.$$

Find the centroid $(\bar{x}, \bar{y})$ of V.

Problem A-2

For positive real x, let

$$B_n(x) = 1^x + 2^x + 3^x + \cdots + n^x.$$

Prove or disprove the convergence of

$$\sum_{n=2}^{\infty} \frac{B_n(\log_n 2)}{(n \log_2 n)^2}.$$

Problem A-3

Evaluate

$$\int_0^{\infty} \frac{\operatorname{Arctan}(\pi x) - \operatorname{Arctan} x}{x} \, dx.$$

Problem A-4

Assume that the system of simultaneous differential equations

$$y' = -z^3, \quad z' = y^3$$

with the initial conditions $y(0) = 1$, $z(0) = 0$ has a unique solution $y = f(x)$, $z = g(x)$ defined for all real x. Prove that there exists a positive constant L such that for all real x,

$$f(x + L) = f(x), \quad g(x + L) = g(x).$$

Problem A-5

Let a, b, c, and d be positive integers and

$$r = 1 - \frac{a}{b} - \frac{c}{d}.$$

Given that $a + c \leqslant 1982$ and $r > 0$, prove that

$$r > \frac{1}{1983^3}.$$

Problem A-6

Let σ be a bijection of the positive integers, that is, a one-to-one function from $\{1, 2, 3, \dots\}$ onto itself. Let $x_1, x_2, x_3, \dots$ be a sequence of real numbers with the following three properties:

(i) $|x_n|$ is a strictly decreasing function of n;
(ii) $|\sigma(n) - n| \cdot |x_n| \to 0$ as $n \to \infty$;
(iii) $\lim_{n\to\infty} \sum_{k=1}^{n} x_k = 1$.

Prove or disprove that these conditions imply that

$$\lim_{n\to\infty} \sum_{k=1}^{n} x_{\sigma(k)} = 1.$$

Problem B-1

Let M be the midpoint of side BC of a general ΔABC. Using the *smallest possible* n, describe a method for cutting ΔAMB into n triangles which can be reassembled to form a triangle congruent to ΔAMC.

Problem B-2

Let $A(x, y)$ denote the number of points (m, n) in the plane with integer coordinates m and n satisfying $m^2 + n^2 \leqslant x^2 + y^2$. Let $g = \sum_{k=0}^{\infty} e^{-k^2}$. Express

$$\int_{-\infty}^{\infty} \int_{-\infty}^{\infty} A(x, y) e^{-x^2-y^2}\, dx\, dy$$

as a polynomial in g.

Problem B-3

Let p_n be the probability that $c + d$ is a perfect square when the integers c and d are selected independently at random from the set $\{1, 2, 3, \ldots, n\}$. Show that $\lim_{n\to\infty}(p_n\sqrt{n})$ exists and express this limit in the form $r(\sqrt{s} - t)$, where s and t are integers and r is a rational number.

Problem B-4

Let $n_1, n_2, \ldots, n_s$ be distinct integers such that

$$(n_1 + k)(n_2 + k) \cdots (n_s + k)$$

is an integral multiple of $n_1 n_2 \cdots n_s$ for every integer k. For each of the following assertions, give a proof or a counterexample:

(a) $|n_i| = 1$ for some i.
(b) If further all n_i are positive, then

$$\{n_1, n_2, \ldots, n_s\} = \{1, 2, \ldots, s\}.$$

Problem B-5

For each $x > e^e$ define a sequence $S_x = u_0, u_1, u_2, \ldots$ recursively as follows: $u_0 = e$, while for $n \geqslant 0$, u_{n+1} is the logarithm of x to the base u_n. Prove that S_x converges to a number $g(x)$ and that the function g defined in this way is continuous for $x > e^e$.

Problem B-6

Let $K(x, y, z)$ denote the area of a triangle whose sides have lengths x, y, and z. For any two triangles with sides a, b, c and a', b', c', respectively, prove that

$$\sqrt{K(a, b, c)} + \sqrt{K(a', b', c')} \leqslant \sqrt{K(a + a', b + b', c + c')}$$

and determine the cases of equality.

THE FORTY-FOURTH WILLIAM LOWELL PUTNAM MATHEMATICAL COMPETITION

December 3, 1983

Problem A-1

How many positive integers n are there such that n is an exact divisor of at least one of the numbers $10^{40}, 20^{30}$?

Problem A-2

The hands of an accurate clock have lengths 3 and 4. Find the distance between the tips of the hands when that distance is increasing most rapidly.

Problem A-3

Let p be in the set $\{3, 5, 7, 11, \dots\}$ of odd primes and let

$$F(n) = 1 + 2n + 3n^2 + \cdots + (p-1)n^{p-2}.$$

Prove that if a and b are distinct integers in $\{0, 1, 2, \dots, p-1\}$ then $F(a)$ and $F(b)$ are not congruent modulo p, that is, $F(a) - F(b)$ is not exactly divisible by p.

Problem A-4

Let k be a positive integer and let $m = 6k - 1$. Let

$$S(m) = \sum_{j=1}^{2k-1} (-1)^{j+1} \binom{m}{3j-1}.$$

For example with $k = 3$,

$$S(17) = \binom{17}{2} - \binom{17}{5} + \binom{17}{8} - \binom{17}{11} + \binom{17}{14}.$$

Prove that $S(m)$ is never zero. $\left[\text{As usual, } \binom{m}{r} = \dfrac{m!}{r!(m-r)!}.\right]$

Problem A-5

Prove or disprove that there exists a positive real number u such that $[u^n] - n$ is an even integer for all positive integers n.

Here $[x]$ denotes the greatest integer less than or equal to x.

Problem A-6

Let $\exp(t)$ denote e^t and

$$F(x) = \frac{x^4}{\exp(x^3)} \int_0^x \int_0^{x-u} \exp(u^3 + v^3)\, dv\, du.$$

Find $\lim_{x \to \infty} F(x)$ or prove that it does not exist.

Problem B-1

Let v be a vertex (corner) of a cube C with edges of length 4. Let S be the largest sphere that can be inscribed in C. Let R be the region consisting of all points p between S and C such that p is closer to v than to any other vertex of the cube. Find the volume of R.

Problem B-2

For positive integers n, let $C(n)$ be the number of representations of n as a sum of nonincreasing powers of 2, where no power can be used more than three times. For example, $C(8) = 5$ since the representations for 8 are:

$$8,\quad 4+4,\quad 4+2+2,\quad 4+2+1+1,\quad \text{and}\quad 2+2+2+1+1.$$

Prove or disprove that there is a polynomial $P(x)$ such that $C(n) = [P(n)]$ for all positive integers n; here $[u]$ denotes the greatest integer less than or equal to u.

Problem B-3

Assume that the differential equation

$$y''' + p(x)y'' + q(x)y' + r(x)y = 0$$

has solutions $y_1(x)$, $y_2(x)$, and $y_3(x)$ on the whole real line such that

$$y_1^2(x) + y_2^2(x) + y_3^2(x) = 1$$

for all real x. Let

$$f(x) = (y_1'(x))^2 + (y_2'(x))^2 + (y_3'(x))^2.$$

Find constants A and B such that $f(x)$ is a solution to the differential equation

$$y' + Ap(x)y = Br(x).$$

Problem B-4

Let $f(n) = n + [\sqrt{n}]$ where $[x]$ is the largest integer less than or equal to x. Prove that, for every positive integer m, the sequence

$$m, f(m), f(f(m)), f(f(f(m))), \ldots$$

contains at least one square of an integer.

Problem B-5

Let $\|u\|$ denote the distance from the real number u to the nearest integer. (For example, $\|2.8\| = .2 = \|3.2\|$.) For positive integers n, let

$$a_n = \frac{1}{n}\int_1^n \left\|\frac{n}{x}\right\| dx.$$

Determine $\lim_{n\to\infty} a_n$. You may assume the identity

$$\frac{2}{1}\cdot\frac{2}{3}\cdot\frac{4}{3}\cdot\frac{4}{5}\cdot\frac{6}{5}\cdot\frac{6}{7}\cdot\frac{8}{7}\cdot\frac{8}{9}\cdot\cdots = \frac{\pi}{2}.$$

Problem B-6

Let k be a positive integer, let $m = 2^k + 1$, and let $r \neq 1$ be a complex root of $z^m - 1 = 0$. Prove that there exist polynomials $P(z)$ and $Q(z)$ with integer coefficients such that

$$(P(r))^2 + (Q(r))^2 = -1.$$

THE FORTY-FIFTH WILLIAM LOWELL PUTNAM MATHEMATICAL COMPETITION

December 1, 1984

Problem A-1

Let A be a solid $a \times b \times c$ rectangular brick in three dimensions, where $a, b, c > 0$. Let B be the set of all points which are a distance at most one from some point of A (in particular, B contains A). Express the volume of B as a polynomial in a, b, and c.

Problem A-2

Express $\sum_{k=1}^{\infty}(6^k/(3^{k+1}-2^{k+1})(3^k-2^k))$ as a rational number.

Problem A-3

Let n be a positive integer. Let a, b, x be real numbers, with $a \neq b$, and let M_n denote the $2n \times 2n$ matrix whose (i, j) entry m_{ij} is given by

$$m_{ij} = \begin{cases} x & \text{if } i = j, \\ a & \text{if } i \neq j \quad \text{and} \quad i + j \text{ is even}, \\ b & \text{if } i \neq j \quad \text{and} \quad i + j \text{ is odd}. \end{cases}$$

Thus, for example, $M_2 = \begin{pmatrix} x & b & a & b \\ b & x & b & a \\ a & b & x & b \\ b & a & b & x \end{pmatrix}$. Express $\lim_{x \to a} \det M_n/(x-a)^{2n-2}$ as a polynomial in a, b, and n, where $\det M_n$ denotes the determinant of M_n.

Problem A-4

A convex pentagon $P = ABCDE$, with vertices labeled consecutively, is inscribed in a circle of radius 1. Find the maximum area of P subject to the condition that the chords AC and BD be perpendicular.

Problem A-5

Let R be the region consisting of all triples (x, y, z) of nonnegative real numbers satisfying $x + y + z \leqslant 1$. Let $w = 1 - x - y - z$. Express the value of the triple integral

$$\iiint_R x^1 y^9 z^8 w^4 \, dx \, dy \, dz$$

in the form $a!b!c!d!/n!$, where a, b, c, d, and n are positive integers.

Problem A-6. Let n be a positive integer, and let $f(n)$ denote the last nonzero digit in the decimal expansion of $n!$. For instance, $f(5) = 2$.

(a) Show that if $a_1, a_2, \ldots, a_k$ are *distinct* nonnegative integers, then $f(5^{a_1} + 5^{a_2} + \cdots + 5^{a_k})$ depends only on the sum $a_1 + a_2 + \cdots + a_k$.

(b) Assuming part (a), we can define

$$g(s) = f(5^{a_1} + 5^{a_2} + \cdots + 5^{a_k}),$$

where $s = a_1 + a_2 + \cdots + a_k$. Find the least positive integer p for which

$$g(s) = g(s + p), \quad \text{for all } s \geqslant 1,$$

or else show that no such p exists.

Problem B-1

Let n be a positive integer, and define

$$f(n) = 1! + 2! + \cdots + n!.$$

Find polynomials $P(x)$ and $Q(x)$ such that

$$f(n+2) = P(n)f(n+1) + Q(n)f(n),$$

for all $n \geqslant 1$.

Problem B-2. Find the minimum value of

$$(u-v)^2 + \left(\sqrt{2-u^2} - \frac{9}{v}\right)^2$$

for $0 < u < \sqrt{2}$ and $v > 0$.

Problem B-3

Prove or disprove the following statement: If F is a finite set with two or more elements, then there exists a binary operation $*$ on F such that for all x, y, z in F,

(i) $x * z = y * z$ implies $x = y$ (right cancellation holds),

and

(ii) $x * (y * z) \neq (x * y) * z$ (*no* case of associativity holds).

Problem B-4

Find, with proof, all real-valued functions $y = g(x)$ defined and *continuous* on $[0, \infty)$, positive on $(0, \infty)$, such that for all $x > 0$ the y-coordinate of the centroid of the region

$$R_x = \{(s,t) | 0 \leqslant s \leqslant x, \quad 0 \leqslant t \leqslant g(s)\}$$

is the same as the average value of g on $[0, x]$.

Problem B-5

For each nonnegative integer k, let $d(k)$ denote the number of 1's in the binary expansion of k (for example, $d(0) = 0$ and $d(5) = 2$). Let m be a positive integer. Express

$$\sum_{k=0}^{2^m - 1} (-1)^{d(k)} k^m$$

in the form $(-1)^m a^{f(m)} (g(m))!$, where a is an integer and f and g are polynomials.

Problem B-6

A sequence of convex polygons $\{P_n\}$, $n \geqslant 0$, is defined inductively as follows. P_0 is an equilateral triangle with sides of length 1. Once P_n has been determined, its sides are trisected; the vertices of P_{n+1} are the *interior* trisection points of the sides of P_n. Thus, P_{n+1} is obtained by cutting corners off P_n, and P_n has $3 \cdot 2^n$ sides. (P_1 is a regular hexagon with sides of length $1/3$.)

Express $\lim_{n \to \infty} \text{Area}(P_n)$ in the form $\sqrt{a}/b$, where a and b are positive integers.

SOLUTIONS

THE TWENTY-SIXTH WILLIAM LOWELL PUTNAM MATHEMATICAL COMPETITION

November 20, 1965

A-1. Suppose the bisector of the exterior angle at A intersects line BC at X and the bisector of the exterior angle at B meets the line AC at Y. The assumption that C is between B and X contradicts the fact that $\angle B > \angle C$ so we may assume that B is between X and C. Similarly, we conclude that C is between A and Y because $\angle A < \angle C$.

If Z is a point on line AB with B between A and Z, we have from triangle ABY that $\angle ZBY = 2A$. Hence, $\angle BXA = \angle ABX = \angle ZBC = 2\angle ZBY = 4A$, and the angle sum of triangle ABX is $90° - \frac{1}{2}A + 8A$. Thus, $A = 12°$.

A-2. Substituting $s = n - r$ in the given summation reveals that twice this sum is equal to:

$$\sum_{r=0}^{n}\left\{\frac{n-2r}{n}\binom{n}{r}\right\}^2$$

$$= \sum\left(1 - 2\frac{r}{n}\right)^2\binom{n}{r}^2 = \sum\binom{n}{r}^2 - 4\sum\frac{r}{n}\binom{n}{r}\binom{n}{r} + 4\sum\left(\frac{r}{n}\right)^2\binom{n}{r}^2$$

$$= \binom{2n}{n} - 4\sum\binom{n-1}{r-1}\binom{n}{r} + 4\sum\binom{n-1}{r-1}^2$$

$$= \binom{2n}{n} - 4\binom{2n-1}{n-1} + 4\binom{2n-2}{n-1} = \binom{2n}{n} - 4\binom{2n-2}{n-2}$$

$$= \left\{\frac{2n(2n-1)}{n^2} - 4\frac{n-1}{n}\right\}\binom{2n-2}{n-1} = \frac{2}{n}\binom{2n-2}{n-1}.$$

Comment: This solution assumes the well-known identities

$$\sum\binom{n}{r}^2 = \binom{2n}{n} \quad\text{and}\quad \sum_{r=0}^{k}\binom{m}{k-r}\binom{n}{r} = \binom{m+n}{k}$$

which may be proved by comparing coefficients in the expansion of

$$(1+x)^m \cdot (1+x)^n = (1+x)^{m+n}.$$

A-3. That (A) implies (B) follows from the fact that subsequences of a convergent sequence converge to the limit of the sequence. We simplify the notation by setting $c_r = \exp ia_r$ and $S(t) = c_1 + c_2 + \cdots + c_t$. Note that $|c_r| = 1$ and $|S(t+k) - S(t)| \leqq k$. Suppose now that (B) holds and write $m = n^2 + k$, where $0 \leqq k \leqq 2n$.

$$\left|\frac{S(m)}{m} - \frac{S(n^2)}{n^2}\right| \leqq \left|\frac{S(m)}{m} - \frac{S(n^2)}{m}\right| + \left|\frac{S(n^2)}{n^2} - \frac{S(n^2)}{m}\right|$$

$$\leqq \frac{k}{m} + n^2\left(\frac{1}{n^2} - \frac{1}{m}\right) = \frac{k + m - n^2}{m} = \frac{2k}{m} \leqq \frac{4n}{n^2}.$$

We conclude that $\lim_{m\to\infty}(S(m)/m - S(n^2)/n^2) = 0$ or that $S(m)/m$ converges to α.

A-4. Let b be a boy who dances with a maximal number of girls (i.e., there may be another boy who dances with the same number of girls, but none dances with a greater number). Let g' be a girl with whom b does not dance, and b' a boy with whom g' dances. Among the partners of b, there must be at least one girl g who does not dance with b' (for otherwise b' would have more partners than b). The couples gb and $g'b'$ solve the problem.

A-5. On the basis of the first few cases we conjecture the answer is 2^{n-1} and proceed by induction.

We first show (also by induction) that an n-arrangement ends in 1 or n. Note that when n is deleted from an n-arrangement, the result is an $(n-1)$-arrangement. If an n-arrangement does not end in 1 or n, deletion of n produces an $(n-1)$-arrangement ending (by induction) in $(n-1)$. This implies the n-arrangement ended in n because n cannot precede $(n-1)$.

For any n-arrangement $(a_1, a_2, \cdots, a_n)$ there is another n-arrangement $(a_1', a_2', \cdots, a_n')$, where $a_i' = n+1-a_i$. If one of these ends in n, the other ends in 1 and consequently exactly half of the n-arrangements end in n.

All of the n-arrangements which end in n can be obtained by adjoining an n to the end of all $(n-1)$-arrangements, and by induction there are 2^{n-2} of these. Hence, there are 2^{n-1} n-arrangements.

A-6. The problem is not well set being true only under rather heavy restrictions on the x, y, u and v. For example, all is in order if they are nonnegative. However, if m is rational with odd numerator and odd denominator there are tangent lines to the curve for which $u^n + v^n > 1$, while if m is rational with odd numerator and even denominator then n is rational with odd numerator and odd denominator and there are solutions (u, v) of $u^n + v^n = 1$ such that the line $ux + vy = 1$ is not tangent to the curve $x^m + y^m = 1$.

Let (x_0, y_0) be a point on the curve $x^m + y^m = 1$. The tangent to this curve is $x_0^{m-1}x + y_0^{m-1}y = 1$. If this line is $ux + vy = 1$, then $u = x_0^{m-1}$ and $v = y_0^{m-1}$ with both u and v nonnegative. The relation $1/m + 1/n = 1$ gives $m/(m-1) = n$ and we obtain $u^n + v^n = x_0^m + y_0^m = 1$.

Conversely, let $m^{-1} + n^{-1} = 1$ and let u and v be nonnegative and such that $u^n + v^n = 1$. Define x_0 and y_0 by the equations $x_0 = u^{n/m}$ and $y_0 = v^{n/m}$. Then x_0 and y_0 are nonnegative and $x_0^m + y_0^m = u^n + v^n = 1$. Thus, (x_0, y_0) is on the curve

$x^m+y^m=1$ and the line $ux+vy=1$ is the tangent to the curve by the calculation above.

In this solution we use the fact that for nonnegative a and positive r and s $(a^r)^s=a^{rs}$.

B-1. The change of variables $x_k \to 1-x_k$ yields

$$\int_0^1\int_0^1\cdots\int_0^1 \cos^2\left\{\frac{\pi}{2n}(x_1+x_2+\cdots+x_n)\right\}dx_1dx_2\cdots dx_n$$

$$=\int_0^1\int_0^1\cdots\int_0^1 \sin^2\left\{\frac{\pi}{2n}(x_1+x_2+\cdots+x_n)\right\}dx_1dx_2\cdots dx_n.$$

Each of these expressions, being equal to half their sum, must equal $\frac{1}{2}$. The limit is also $\frac{1}{2}$.

B-2. Clearly $\omega_r+l_r=n-1$ for $r=1, 2, \cdots, n$ and $\sum_1^n \omega_r=\sum_1^n l_r$. Hence,

$$\sum_1^n \omega_r^2-\sum_1^n l_r^2=\sum_1^n(\omega_r-l_r)(\omega_r+l_r)=(n-1)\sum_1^n(\omega_r-l_r)=(n-1)\cdot 0=0.$$

B-3. All Pythagorean triples can be obtained from $x=\lambda(p^2-q^2)$, $y=2\lambda pq$, $z=\lambda(p^2+q^2)$ where $0<q<p$, $(p, q)=1$ and $p\not\equiv q \mod 2$, λ being any natural number.

The problem requires that $\frac{1}{2}xy=2(x+y+z)$. This condition can be written $\lambda^2(p^2-q^2)(pq)=2\lambda(p^2-q^2+2pq+p^2+q^2)$ or simply $\lambda(p-q)q=4$. Since $p-q$ is odd it follows that $p-q=1$ and the only possibilities for q are 1, 2, 4.

If $q=1$, $p=2$, $\lambda=4$, $x=12$, $y=16$, $z=20$.
If $q=2$, $p=3$, $\lambda=2$, $x=10$, $y=24$, $z=26$.
If $q=4$, $p=5$, $\lambda=1$, $x=9$, $y=40$, $z=41$.

B-4. Since

$$\binom{n+1}{r}=\binom{n}{r}+\binom{n}{r-1}, \qquad f(x, n+1)=\frac{f(x, n)+x}{f(x, n)+1}.$$

If x is such that $f(x, n)$ converges when n tends to infinity, the limit $F(x)$ must satisfy $F(x)=(F(x)+x)/(F(x)+1)$, $F^2(x)=x$. The convergence to $\sqrt{x}$ is obvious when $x=0$ or 1. To show this convergence for any positive x we first note that

$$f(x, n)=\sqrt{x}\,\frac{(1+\sqrt{x})^n+(1-\sqrt{x})^n}{(1+\sqrt{x})^n-(1-\sqrt{x})^n}.$$

When $0<x<1$, write $a=(1-\sqrt{x})/(1+\sqrt{x})$; then $0<a<1$ and

$$f(x, n)=\sqrt{x}\,\frac{1+a^n}{1-a^n}\to\sqrt{x}.$$

When $x>1$, write $b=(\sqrt{x}-1)/(\sqrt{x}+1)$; then $0<b<1$ and

$$f(x, n) = \sqrt{x}\frac{1+(-b)^n}{1-(-b)^n} \to \sqrt{x}.$$

The limit fails to exist for negative values of x; but for all other complex numbers the limit exists and is that square root of x which lies in the right half plane.

B-5. Divide the objects into two subsets $\{a_1, a_2, \cdots, a_m\}$ and $\{b_1, b_2, \cdots, b_n\}$, where $m+n=V$. Then the mn pairs (a_j, b_k), where $j=1, 2, \cdots, m$ and $k=1, 2, \cdots, n$, obviously contain no triangles. If V is even, take $m=n=V/2$, and if V is odd, take $m=(V+1)/2$, $n=(V-1)/2$. Then $mn \geqq V^2/4 \geqq E$.

B-6. Suppose A, B, C, D are neither concyclic nor collinear. Then p, the perpendicular bisector of segment AB, cannot coincide with q, the perpendicular bisector of segment CD. If the lines p and q intersect, their common point is the center of two *concentric* circles, one through A and B, the other through C and D. If instead p and q are parallel, so also are the lines AB and CD. Consider points P and Q, on p and q respectively, midway between the parallel lines AB and CD. Clearly, the circles ABP and CDQ have no common point.

THE TWENTY-SEVENTH WILLIAM LOWELL PUTNAM MATHEMATICAL COMPETITION

November 19, 1966

A-1 It is easily verified by induction that

$$f(n) = \begin{cases} n^2/4 & \text{when } n \text{ is even,} \\ (n^2-1)/4 & \text{when } n \text{ is odd.} \end{cases}$$

Therefore, since $x+y$ and $x-y$ always have the same parity, in any case we must have

$$f(x+y) - f(x-y) = \frac{(x+y)^2 - (x-y)^2}{4} = xy.$$

A-2 The area of the given triangle can be calculated in two ways, $A = pr$ and $A = \sqrt{p(p-a)(p-b)(p-c)}$. Squaring and equating we get $p^2r^2 = p(p-a)(p-b)(p-c)$. Setting $x = 1/(p-a)$, $y = 1/(p-b)$, $z = 1/(p-c)$ we can write this equation in the form

$$\frac{1}{r^2} = pxyz = xyz\left(\frac{1}{x} + \frac{1}{y} + \frac{1}{z}\right).$$

Thus we need only show that $yz + xz + xy \leqq x^2 + y^2 + z^2$. However this follows from the trivial inequalities $y^2 + z^2 \geqq 2yz$, $x^2 + z^2 \geqq 2xz$, $x^2 + y^2 \geqq 2xy$.

A-3 Multiplying the defining relation by $(n+1)$ we get

(1) $$(n+1)x_{n+1} = nx_n + x_n - (n+1)(x_n)^2 = nx_n + x_n[1 - (n+1)x_n].$$

To prove that nx_n is increasing, we need to show that $1 - (n+1)x_n \geqq 0$. From the graph of $x(1-x)$ we note that $x_2 \leqq \frac{1}{4}$ and that $x_n \leqq a \leqq \frac{1}{2}$ implies $x_{n+1} \leqq a(1-a)$. So by induction,

$$(n+1)x_n \leqq (n+1)\frac{1}{n}\left(1 - \frac{1}{n}\right) = 1 - \frac{1}{n^2} \leqq 1.$$

Furthermore, $nx_n < (n+1)x_n \leqq 1$ and so nx_n is bounded above by 1. Thus nx_n converges to a limit α with $0 < nx_n < \alpha \leqq 1$. Now summing (1) from 2 to n we obtain

$$\begin{aligned}(2)\qquad 1 &\geqq (n+1)x_{n+1}\\ &= 2x_2 + x_2(1-3x_2) + x_3(1-4x_3) + \cdots + x_n[1-(n+1)x_n].\end{aligned}$$

If $\alpha \neq 1$ then $[1-(n+1)x_n] \geqq (1-\alpha)/2$ for all large n and thus (2) would show that $\sum x_n$ is convergent. However $nx_n \geqq x_1$ and so $\sum x_n \geqq x_1 \sum (1/n)$.

A-4 To prove the formula by induction, it suffices to show that the difference $\Delta = n + \{\sqrt{n}\} - (n-1+\{\sqrt{n-1}\}) = 1$ or 2, with the value 2 occurring if and only if the number $n+\{\sqrt{n-1}\}$ is a perfect square. For convenience, let $\{\sqrt{n-1}\} = q$. Then of course $q-\frac{1}{2} < \sqrt{n-1} < q+\frac{1}{2}$ or better $q^2-q+\frac{1}{4} < n-1 < q^2+q+\frac{1}{4}$. This gives $q^2+5/4 < n+\{\sqrt{n-1}\} < (q+1)^2+\frac{1}{4}$. Therefore the number $n+\{\sqrt{n-1}\}$ is a perfect square if and only if $n=(q+1)^2-q$. However, then and only then $\sqrt{n} > q+\frac{1}{2} > \sqrt{n-1}$. In other words then and only then $\{\sqrt{n}\} - \{\sqrt{n-1}\} = 1$, because this difference is never greater than 1.

A-5 We shall show that $T\psi(x) = f(x)\psi(x)$ for all $\psi \in C$, where f is the image under T of the function 1 which sends x into 1. For each x_0 and ψ, define ψ' by

$$\psi'(x) = \begin{cases} \psi(x) & \text{if } x \leqq x_0, \\ \psi(x_0) & \text{if } x > x_0. \end{cases}$$

Since T is local we must have $T\psi'(x) = T\psi(x)$ for all $x \leqq x_0$. On the other hand, for $x > x_0$, $T\psi'(x) = \psi(x_0)T1(x) = \psi(x_0)\cdot f(x)$. By continuity of $T\psi'$, $T\psi(x_0) = \psi(x_0)\cdot f(x_0)$.

Comment: The condition (1) is needed only in the case where $c_2 = 0$. Also the interval I in condition (2) should be required to have positive measures since the problem is trivial if I can be taken as a point.

A-6 We understand the statement to mean that

$$3 = \lim_{n\to\infty} \sqrt{1+2\sqrt{1+3\sqrt{1+\cdots\sqrt{1+(n-1)\sqrt{1+n}}}}}\,.$$

We see that

$$\begin{aligned}3 &= \sqrt{1+2\cdot 4} = \sqrt{1+2\sqrt{16}} = \sqrt{1+2\sqrt{1+3\sqrt{25}}}\\ &= \sqrt{1+2\sqrt{1+3\sqrt{1+4\sqrt{36}}}}.\end{aligned}$$

This leads us to conjecture the relation

$$3 = \sqrt{1+2\sqrt{1+3\sqrt{1+\cdots+\sqrt{1+n\sqrt{(n+2)^2}}}}} \qquad \text{for all } n \geqq 1.$$

Proceeding by induction we verify that $(n+2)^2 = n^2+4n+4 = 1+(n+1)(n+3) = 1+(n+1)\sqrt{(n+3)^2}$. This given, we must have

$$3 \geqq \sqrt{1+2\sqrt{1+3\sqrt{\cdots\sqrt{1+(n-1)\sqrt{(1+n)}}}}}.$$

To set an inequality in the other direction, observe that for any $\alpha > 1$

$$\sqrt{1+n\alpha} \leqq \sqrt{\alpha}\sqrt{1+n}.$$

A repetition of this inequality gives then

$3 \leqq (n+2)^a \sqrt{1+2\sqrt{1+3\sqrt{1+\cdots\sqrt{1+(n-1)\sqrt{1+n}}}}}$, where $a = 2^{1-n}$.

B-1 Let $P_1, P_2, \cdots, P_n$ be the vertices of P. Let $P_1', P_2', \cdots, P_n'$ be the projections of $P_1, P_2, \cdots, P_n$ upon one of the sides of the squares, and let $P_1'', P_2'', \cdots, P_n''$ be the projections of $P_1, P_2, \cdots, P_n$ upon a side that is orthogonal to the previous one. Since P is convex, the first side will be covered at most twice by the segments $\overline{P_1'P_2'}, \cdots, \overline{P_{n-1}'P_n'}, \overline{P_n'P_1'}$. We thus deduce the inequality $\overline{P_1'P_2'}^2 + \cdots + \overline{P_n'P_1'}^2 \leqq 2$. Similarly $\overline{P_1''P_2''}^2 + \cdots + \overline{P_n''P_1''}^2 \leqq 2$. Adding these two inequalities and using the Pythagorean theorem the assertion follows.

B-2 Any common factor of two of such numbers would have to be divisible by 2, 3, 5 or 7. So it is sufficient to prove that among any ten consecutive integers there is at least one that is not divisible by 2, 3, 5 or 7. We get such an integer by elimination as follows. Strike out those divisible by 3. There may be either 3 or 4 of them. Among these there is either at least one or two respectively that are divisible also by 2. Thus if we strike off also those that are divisible by 2 we will have eliminated at most seven of the integers. Note that by so doing we have stricken off at least one number divisible by five. Thus we are left with three integers only two of which can be divisible by 5 or 7.

B-3 Set $q_n = p_1 + p_2 + \cdots + p_n (q_0 = 0)$. We are led to estimate $S_N = \sum_{n=1}^{N} (n/q_n)^2 (q_n - q_{n-1})$ in terms of $T = \sum_{n=1}^{\infty} 1/p_n$. Note that

$$S_N \leqq \frac{1}{p_1} + \sum_{n=2}^{N} \frac{n^2}{q_n q_{n-1}} (q_n - q_{n-1}) = \frac{1}{p_1} + \sum_{n=2}^{N} \frac{n^2}{q_{n-1}} - \sum_{n=2}^{N} \frac{n^2}{q_n}$$

$$= \frac{1}{p_1} + \sum_{n=1}^{N-1} \frac{(n+1)^2}{q_n} - \sum_{n=2}^{N} \frac{n^2}{q_n} \leqq \frac{5}{p_1} + 2\sum_{n=2}^{N} \frac{n}{q_n} + \sum_{n=2}^{N} \frac{1}{q_n}.$$

By Schwarz's inequality,

$$\left(\sum_{n=2}^{N} \frac{n}{q_n}\right)^2 \leqq \sum_{n=2}^{N} \frac{n^2}{q_n^2} p_n \sum_{n=1}^{\infty} \frac{1}{p_n}$$

and thus

$$S_N \leqq \frac{5}{p_1} + 2\sqrt{S_N T} + T.$$

This quadratic inequality implies that $\sqrt{S_N} \leqq \sqrt{T} + \sqrt{2T + 5/p_1}$.

B-4 Let, for each $1 \leqq i \leqq mn+1$, n_i denote the length of the longest chain, starting with a_i and each dividing the following one, that we can select out of $a_i, a_{i+1}, \cdots, a_{mn+1}$. If no n_i is greater than n then there are at least $m+1$ n_i's that are the same. However, the integers a_i corresponding to these n_i's cannot divide each other, because $a_i | a_j$ implies that $n_i \geqq n_j + 1$.

B-5 Let these points be denoted by $P_1, P_2, \cdots, P_n$. To every permutation $(\sigma_1, \sigma_2, \cdots, \sigma_n)$ of $(1, 2, 3, \cdots, n)$ we associate a closed polygon, namely $P_{\sigma_1}P_{\sigma_2} \cdots P_{\sigma_n}P_{\sigma_1}$. This way we obtain $(n-1)!$ distinct closed polygons some of which may have selfintersections. We claim that anyone of these polygons whose length is the shortest possible is simple. By the hypothesis that no three P_i's are on the same line, a selfintersection occurs if and only if two segments say $\overline{P_{\sigma_1}P_{\sigma_2}}$ and $\overline{P_{\sigma_m}P_{\sigma_{m+1}}}$ cross each other. However, then the closed polygon $P_{\sigma_2} \cdots P_{\sigma_{m-1}}P_{\sigma_m}P_{\sigma_1}P_{\sigma_n}P_{\sigma_{n-1}} \cdots P_{\sigma_{m+1}}P_{\sigma_2}$ would have shorter length. Thus there can't be a cross if the length of $P_{\sigma_1} \cdots P_{\sigma_n}P_{\sigma_1}$ is shortest possible.

Alternate solution: Take two points P_1 and P_2 such that all the other points $P_3, P_4, \cdots, P_n$ are on the same side of the line connecting P_1 and P_2. Each point P_i, $i>2$, determines an angle θ_i between P_1P_2 and P_1P_i, with $0<\theta_i<\pi$. By hypothesis, $\theta_i \neq \theta_j$ if $i \neq j$. Let $(i_3, i_4, \cdots, i_n)$ be the permutation of $(3, 4, \cdots, n)$ such that $\theta_{i_3}<\theta_{i_4}< \cdots <\theta_{i_n}$. Then $P_1P_2P_{i_3}P_{i_4} \cdots P_{i_n}P_1$ is a closed simple polygon.

B-6 Multiplying the equation by $e^{-x}y'$ and integrating from 0 to T we get, after rearranging terms,

$$y^2(T) + 2\int_0^T e^{-x}y'y''\,dx = y^2(0).$$

Note then that for some $0 \leq \zeta \leq T$ we must have

$$2\int_0^T e^{-x}y'y''\,dx = 2\int_0^\zeta y'y''\,dx = \{y'(\zeta)\}^2 - \{y'(0)\}^2.$$

We then obtain $y^2(T) + \{y'(\zeta)\}^2 = y^2(0) + \{y'(0)\}^2$.

THE TWENTY-EIGHTH WILLIAM LOWELL PUTNAM MATHEMATICAL COMPETITION

December 2, 1967

A-1

$$|a_1 + 2a_2 + \cdots + na_n| = |f'(0)| = \lim_{x\to 0}\left|\frac{f(x) - f(0)}{x}\right|$$

$$= \lim_{x\to 0}\left|\frac{f(x)}{\sin x}\right|\cdot\left|\frac{\sin x}{x}\right| = \lim_{x\to 0}\left|\frac{f(x)}{\sin x}\right| \leqq 1.$$

A-2 S_n is the number of symmetric $n\times n$ permutation matrices (a permutation matrix has exactly one 1 in each row and column with 0's elsewhere). Let the 1 in the first row be in the kth column. If $k=1$, then there are S_{n-1} ways to complete the matrix. If $k\neq 1$ then $a_{1k}=a_{k1}=1$ and the deletion of the 1st and kth rows and columns leaves a symmetric $(n-2)\times(n-2)$ permutation matrix. Consequently $S_n = S_{n-1}+(n-1)S_{n-2}$.

For part (b), let

$$F(x) = \sum_{n=0}^{\infty}\left\{S_n\frac{x^n}{n!}\right\}. \quad F'(x) = \sum_{n=1}^{\infty}\left\{S_n\frac{x^{n-1}}{(n-1)!}\right\}$$

$$= \sum_{n=1}^{\infty}\left\{S_{n-1}\frac{x^{n-1}}{(n-1)!} + (n-1)S_{n-2}\frac{x^{n-1}}{(n-1)!}\right\}$$

$$= \sum_{n=0}^{\infty}\left\{S_n\frac{x^n}{n!}\right\} + \sum_{n=2}^{\infty}\left\{S_{n-2}\frac{x^{n-1}}{(n-2)!}\right\} = F(x) + xF(x).$$

Hence $F'(x)/F(x) = 1+x$. Integration and use of $F(0)=S_0=1$ yields $F(x) = \exp(x+x^2/2)$. Now the series for $F(x)$ is uniformly convergent for all x, so all the operations are legal.

Comment: S_n is the number of permutations π, with $\pi^2=1$, in the symmetric group on n symbols. Some contestants used this observation together with known formulas for the number of permutations with a given cycle structure to check part (b) by multiplying the series for $\exp(x)$ and $\exp(x^2/2)$.

A-3 Let $f(x)=ax^2-bx+c=a(x-r)(x-s)$. Then $f(0)\cdot f(1)=a^2r(r-1)\cdot s(s-1)$. The graph of $r(r-1)$ shows that $0<r<1$ implies $0<r(r-1)\leqq\frac{1}{4}$, with equality if and only if $r=\frac{1}{2}$. Similarly, $0<s(s-1)\leqq\frac{1}{4}$. Since $r\neq s$, $r(r-1)s(s-1) <1/16$ and $0<f(0)\cdot f(1)<a^2/16$. The coefficients a, b, c are integers and thus

$1 \leqq f(0) \cdot f(1)$. Consequently $a^2 > 16$, i.e. $a \geqq 5$.

The discriminant $b^2 - 4ac$ shows that the minimum possible value for b is 5. Furthermore, $5x^2 - 5x + 1$ has two distinct roots between 0 and 1.

A-4 Assuming that there is a solution u, then integrating with respect to x from 0 to 1, one obtains $\int_0^1 u(x)dx = \int_0^1 1 \cdot dx + \lambda \int_0^1 \{\int_x^1 u(y)u(y-x)dy\} dx$. In the iterated integral, one can interchange the order of integration, and letting $\int_0^1 u(x)dx = \alpha$, one gets $\alpha = 1 + \lambda \int_0^1 \{u(y) \int_0^y u(y-x)dx\} dy$. Now, holding y fixed, let $z = y - x$ to get $\alpha = 1 + \lambda \int_0^1 \{u(y) \int_0^y u(z)dz\} dy$. Set $f(y) = \int_0^y u(z)dz$. Then $\alpha = 1 + \lambda \int_0^1 f'(y)f(y)dy = 1 + \lambda\{\frac{1}{2}f^2(1) - \frac{1}{2}f^2(0)\} = 1 + \lambda \cdot \frac{1}{2}\alpha^2$, or $\lambda\alpha^2 - 2\alpha + 2 = 0$. The discriminant of this quadratic shows that if $\lambda > \frac{1}{2}$ then the roots are imaginary.

A-5 Let the maximum diameter be $2d$ and assume $d < \frac{1}{2}$. Take such a diameter as the x-axis with the origin at the mid-point. Since this is a maximum diameter the region is bounded between the lines $x = -d$ and $x = d$. The upper and lower boundaries of the region are functions, because of convexity. Denote them by $f(x)$ and $-g(x)$, where f and g are nonnegative for $-d \leqq x \leqq d$. Calculating the distance between $(x, f(x))$ and $(-x, -g(x))$ shows that $f(x) + g(-x) < \sqrt{(1+4x^2)}$, for $-d \leqq x \leqq d$. Area $= \int_{-d}^{d} \{f(x) + g(-x)\} dx < \int_{-d}^{d} \sqrt{(1+4x^2)}dx < \int_{-1/2}^{1/2} \sqrt{(1+4x^2)}dx = \frac{1}{4}\pi$. This contradiction proves that $d \geqq \frac{1}{2}$ and so there must be at least two points a unit distance apart.

Comment: The requirement that the boundary contain at most a finite number of straight line segments was extraneous.

A-6 Solving the given equations in terms of x_3 and x_4, leads to the equivalent system: $x_1 = A_1x_3 + B_1x_4$, $x_2 = A_2x_3 + B_2x_4$, $x_3 = x_3$, $x_4 = x_4$, where $A_1 = (a_2b_3 - a_3b_2)/(a_1b_2 - a_2b_1)$, etc.

Each point in the x_3, x_4-plane corresponds uniquely to a solution (x_1, x_2, x_3, x_4). Signum x_1 is positive for (x_3, x_4) on one side of $A_1x_3 + B_1x_4 = 0$ and negative on the other side. Similarly for signum x_2, signum x_3, and signum x_4, using $A_2x_3 + B_2x_3 = 0$, $x_3 = 0$ and $x_4 = 0$, respectively. These four lines through the origin, in general, divide the plane into eight regions, each having a different 4-tuple of signum values. Hence the maximum number of distinct 4-tuples is eight.

The maximum number of eight will occur if and only if there are actually four distinct lines. This is equivalent to the conditions $A_1 \neq 0$, $A_2 \neq 0$, $B_1 \neq 0$, $B_2 \neq 0$ and $A_1B_2 - A_2B_1 \neq 0$ or, simply, $a_ib_j - a_jb_i \neq 0$ for $i, j = 1, 2, 3, 4$ and $i < j$.

B-1 Consider the figure in the complex plane with the center of the circle at the origin. We can take A, B, C, D, E, F as complex numbers of absolute value r. Furthermore $B = A\omega$, $D = C\omega$ and $F = E\omega$, where $\omega = \cos(\pi/3) + i \sin(\pi/3)$. Since $\omega^3 = 1$ and $\omega \neq 1$, $\omega^2 - \omega + 1 = 0$. The mid-points of BC, DE and FA are $P = \frac{1}{2}(A\omega + C)$, $Q = \frac{1}{2}(C\omega + E)$ and $R = \frac{1}{2}(E\omega + A)$. If the segment from Q to R is rotated through $\pi/3$ about Q, then R is carried into $Q + \omega(R - Q)$, which equals P. Thus P, Q, R are vertices of an equilateral triangle.

B-2 For part (a) one has immediately that $A = p^2$, $B = 2p(1-p)$, and

$C=(1-p)^2$. The result follows by examination of the graphs for A, B and C on $0 \leqq p \leqq 1$.

For part (b), $\alpha=pr$, $\beta=p(1-r)+r(1-p)$, $\gamma=(1-p)(1-r)$. Consider the region R in the p, r-plane defined by $0 \leqq p \leqq 1$ and $0 \leqq r \leqq 1$. We will show that there is no point in R with $\alpha<4/9$, $\beta<4/9$ and $\gamma<4/9$. If $\alpha<4/9$ and $\gamma<4/9$ then (p, r) is between the hyperbolas $pr=4/9$ and $(1-p)(1-r)=4/9$. These hyperbolas have vertices in R at $(2/3, 2/3)$ and $(1/3, 1/3)$, respectively. The symmetry about $(\frac{1}{2}, \frac{1}{2})$ suggests setting $p'=p-\frac{1}{2}$ and $r'=r-\frac{1}{2}$. Then $\beta=\frac{1}{2}-2p'r'$ and thus $\beta<4/9$ if and only if $p'r'>1/36$. Note that the vertices for the hyperbola $p'r'=1/36$ are at $(p, r)=(1/3, 1/3)$ and $(2/3, 2/3)$. By looking at asymptotes, we see graphically that the region $\beta<4/9$ does not overlap the region in R where $\alpha<4/9$ and $\gamma<4/9$.

B-3 Since f is uniformly continuous, for any $\epsilon>0$ there is an integer n such that $|x-y|<1/n$ implies $|f(x)-f(y)|<\epsilon$.

$$\int_0^1 f(x)g(nx)dx = \sum_{m=0}^{n-1}\int_{m/n}^{m+1/n} f(x)g(nx)dx = \sum_{m=0}^{n-1}\int_{m/n}^{m+1/n} f(m/n)g(nx)dx$$
$$+\sum_{m=0}^{n-1}\int_{m/n}^{m+1/n}(f(x)-f(m/n))g(nx)dx.$$

The first term equals $\sum_{m=0}^{n-1}(1/n)f(m/n)\int_0^1 g(t)dt$ and becomes $(\int_0^1 f(x)dx)(\int_0^1 g(x)dx)$ as $n\to\infty$. Furthermore,

$$\left|\int_{m/n}^{m+1/n}\{f(x)-f(m/n)\}g(nx)dx\right| < \int_{m/n}^{m+1/n}|f(x)-f(m/n)|\cdot|g(nx)|\,dx$$
$$< \int_{m/n}^{m+1/n}\epsilon|g(nx)|\,dx < \epsilon/n\int_0^1|g(t)|\,dt.$$

Thus the absolute value of the second term is less than or equal to $\epsilon\int_0^1|g(t)|dt$ which becomes 0 as $\epsilon\to 0$.

B-4 Locker m, $1 \leqq m \leqq n$, will be unlocked after the n operations are performed if and only if m has an odd number of positive divisors. If $m=p^\alpha q^\beta \cdots r^\gamma$, then the number of divisors of m is $(\alpha+1)(\beta+1)\cdots(\gamma+1)$, which is odd if and only if $\alpha, \beta, \cdots, \gamma$ are all even. This is equivalent to the condition that m is a perfect square.

For part (b), the set of numbers of the form $2m^2$ are obtained by having T_k change lockers whose numbers are multiples of $2k$. The set m^2+1 results from T_k changing locker i if $i-1$ is a multiple of k, with the stipulation that locker number 1 is changed only by T_1.

B-5 Let A_n be the sum of the first n terms in the binomial expansion of $(2-1)^{-n}$.

$$A_n=\sum_{i=0}^{n-1}\binom{n+i-1}{i}2^{-n-i}=2^{-n}+\sum_{i=1}^{n-1}\left\{\binom{n+i-2}{i}+\binom{n+i-2}{i-1}\right\}2^{-n-i}$$

$$=2^{-n}+\left\{\sum_{i=0}^{n-2}\binom{n+i-2}{i}2^{-n-i}+\binom{2n-3}{n-1}2^{-2n+1}-2^{-n}\right\}+\sum_{j=0}^{n-2}\binom{n+j-1}{j}2^{-n-j-1}$$

$$=2^{-n}+\tfrac{1}{2}A_{n-1}+\binom{2n-3}{n-1}2^{-2n+1}-2^{-n}+\tfrac{1}{2}A_n-\binom{2n-2}{n-1}2^{-2n}$$

$$=\tfrac{1}{2}A_{n-1}+\tfrac{1}{2}A_n+2^{-2n}\left\{2\binom{2n-3}{n-1}-\binom{2n-3}{n-1}-\binom{2n-3}{n-2}\right\}=\tfrac{1}{2}A_{n-1}+\tfrac{1}{2}A_n.$$

Thus $A_n=A_{n-1}$, but $A_1=2^{-1}=\frac{1}{2}$ and so $A_n=\frac{1}{2}$ for all positive integers n.

Alternate solution: Consider a random walk starting at (0, 0), such that if one is at (x, y) the probability of moving to $(x+1, y)$ is $\frac{1}{2}$ and the probability of moving to $(x, y+1)$ is $\frac{1}{2}$. Let S_n be the square with vertices at (0, 0), $(n, 0)$, (n, n), $(0, n)$. By symmetry, the probability $R_i(n)$ of first touching S_n at (n, i), $0\leqq i<n$, equals the probability $T_i(n)$ of first touching S_n at (i, n), $0\leqq i<n$, and hence $\sum_{i=0}^{n-1} R_i(n)=\sum_{i=0}^{n-1} T_i(n)=\frac{1}{2}$. Furthermore, $R_i(n)=\binom{n+i-1}{i}\left(\frac{1}{2}\right)^{n+i}$.

B-6 Consider the function g whose values are defined by $g(x, y)=f(x, y)+2(x^2+y^2)$. On the circumference of the unit circle, $g(x, y)\geqq 1$, and at the origin, $g(0, 0)\leqq 1$. Hence, either $g(x, y)=\text{constant}$ and $f(x, y)=\text{constant}-2(x^2+y^2)$, or there is a minimum value for $g(x, y)$ at some interior point. In the case $g(x, y)=\text{constant}$, the result is immediate. Otherwise, let (x_0, y_0) be the coordinates of a point where $g(x, y)$ has a minimum. Then

$$\frac{\partial g}{\partial x}=\frac{\partial g}{\partial y}=0 \text{ at } (x_0, y_0)$$

$$\text{and}\quad \left|\frac{\partial f}{\partial x}(x_0, y_0)\right|\leqq 4|x_0|,\quad \left|\frac{\partial f}{\partial y}(x_0, y_0)\right|\leqq 4|y_0|.$$

Thus the conclusion follows.

THE TWENTY-NINTH WILLIAM LOWELL PUTNAM MATHEMATICAL COMPETITION

December 7, 1968

A-1 The standard approach, from elementary calculus, applies. By division, rewrite the integrand as a polynomial plus a rational function with numerator of degree less than 2. The solution follows easily.

A-2 The easy solution is obtained by selecting a rational number ρ with $0<\rho<\epsilon$ and solving the linear system

$$ar + bs = e + \rho$$

$$cr + ds = f + \rho.$$

The solution for r and s exist, since $ad \neq bc$, and are rational numbers which satisfy the given inequalities.

A-3 The proof is by induction. For a singleton set $\{1\}$ the list is $\emptyset$, $\{1\}$. Thus the result is true for singleton sets. Suppose the result is true for all sets with $n-1$ members. Let $S=\{1, 2, 3, \cdots, n\}$ and $T=\{1, 2, 3, \cdots, n-1\}$. Let $T_0, T_1, \cdots, T_t$ $(t=2^{n-1}-1)$ be the list of subsets of T satisfying the requirements. Then the desired list of subsets of S are $S_0, S_1, \cdots, S_s$ $(s=2^{s-1})$ where $S_i=T_i$, for $0\leqq i<t$, and $S_t=T_t\cup\{n\}$, $S_{t+1}=T_{t-1}\cup\{n\}, \cdots, S=\{n\}$.

Comments: This problem is equivalent to finding a Hamiltonian circuit on an n-cube.

A-4 The n points can be represented by vectors v_i $(i=1, \cdots, n)$ with $|v_i|=1$. Expanding "the sum of the squares of the distances between them" in the case $n=3$ suggests the following general identities:

$$\begin{aligned}\sum_{1\leq i<j\leq n} |v_i - v_j|^2 &= \sum_{1\leq i<j\leq n} (v_i - v_j)\cdot(v_i - v_j)\\ &= (n-1)\sum_{1\leq i\leq n} v_i\cdot v_i - 2\sum_{1\leq i<j\leq n} v_i\cdot v_j\\ &= n\sum_{1\leq i\leq n} v_i\cdot v_i - \left[\sum_{1\leq i\leq n} v_i\cdot v_i + 2\sum_{1\leq i<j\leq n} v_i\cdot v_j\right]\\ &= n^2 - \left(\sum_{1\leq i\leq n} v_i\right)\cdot\left(\sum_{1\leq i\leq n} v_i\right).\end{aligned}$$

Thus the result follows and, in addition, equality exists if and only if $\sum_{1\leq i\leq n} v_i=0$.

A-5 (11) Let $f(x)=ax^2+bx+c$ be an arbitrary quadratic polynomial. Then $f(0)=c, f(\frac{1}{2})=\frac{1}{4}a+\frac{1}{2}b+c$, and $f(1)=a+b+c. f'(0)=b=4f(\frac{1}{2})-3f(0)-f(1)$. Using the given conditions, $|P'(0)|\leqq 4|P(\frac{1}{2})|+3|P(0)|+|P(1)|\leqq 8$. Furthermore, $P(x)=8x^2-8x+1$ satisfies the given conditions and has $|P'(0)|=8$.

A-6 (0) The desired polynomials with $a_0=-1$ are the negative of those with $a_0=1$, so consider $a_0=1$. The sum of the squares of the zeros of $x^n+a_1x^{n-1}+\cdots+a_n$ is $a_1^2-2a_2$. The product of the squares of these zeros is a_n^2. If all the zeros are real, we can apply the arithmetic-geometric mean inequality to obtain

$$\frac{a_1^2-2a_2}{n}\geqq (a_n^2)^{1/n},$$

with equality only if the zeros are numerically equal. In our case this inequality becomes $(1\pm 2)/n\geqq 1$ or $n\leqq 3$. Note that $n>1$ implies $a_2=-1$ and $n=3$ implies all zeros are ± 1. Thus the list of polynomials is:

$$\pm(x-1), \quad \pm(x+1), \quad \pm(x^2+x-1), \quad \pm(x^2-x-1),$$
$$\pm(x^3+x^2-x-1), \quad \pm(x^3-x^2-x+1).$$

B-1 Denote the four events $x°=70°$, $y°=70°$, $\max(x°, y°)=70°$, $\min(x°, y°)=70°$ by A, B, C, D, respectively. Then $A\cup B=C\cup D$, and $A\cap B=C\cap D$. Hence $P(A)+P(B)=P(A\cup B)+P(A\cap B)=P(C\cup D)+P(C\cap D)=P(C)+P(D)$ and $P(\min(x°, y°)=70°)=P(x°=70°)+P(y°=70°)-P(\max(x°, y°)=70°)$.

B-2 Let g be any element of G. The set $\{ga^{-1}|a\in A\}$ has the same number of elements as A. If these two sets are disjoint, their union would contain more elements than G. Thus there exist $a_1, a_2\in A$ such that $a_1=ga_2^{-1}$ and $g=a_1a_2$.

Alternate Solution: Let G have n elements, A have m elements, and consider the multiplication table of G. An element g in G must appear exactly once in each row and column of the multiplication table. It appears at most $2(n-m)$ times outside the table for A and n times in the table for G. Thus it appears at least $n-2(n-m)=2m-n$ times in the table for A, and we are given that $2m>n$.

B-3 We need to make use of the following facts about fields and constructibility: (1) If Q is the field of rational numbers, the degree of Q extended by $\cos(360°/k)$, where k is a positive integer, is $\phi(k)$, where ϕ is the Euler function. (2) If K, L, M are fields with $K\subset L\subset M$ and $[L:K]<\infty$, $[M:L]<\infty$ then $[M:K]=[M:L]\cdot[L:K]$. (3) Given $\cos(360°/k)$, then $\cos(360°/3k)$ is constructible if and only if

$$\left[Q\left(\cos\frac{360°}{3k}\right):Q\left(\cos\frac{360°}{k}\right)\right]$$

is a power of 2.

Consequently, $[Q(\cos 360°/3k):Q(\cos 360°/k)]\cdot\phi(k)=\phi(3k)$. Now

$$\phi(3^a) = 3^{a-1}\cdot 2 = \begin{cases} 3\phi(3^{a-1}), & \text{if } a > 1 \\ 2, & \text{if } a = 1 \end{cases}$$

and, by the multiplicative property of the Euler function,

$$\phi(3k) = \begin{cases} 3\phi(k), & \text{if } 3 \mid k \\ 2\phi(k), & \text{if } 3 \nmid k. \end{cases}$$

Therefore an angle of size $360°/k$ is trisectible if and only if $3\nmid k$.

B-4 The graph of $y=x-1/x$ suggests splitting the integral into the form

$$\int_{-\infty}^{\infty} f(x-1/x)dx = \lim_{a\to-\infty}\int_a^{-1} f(x-1/x)dx + \lim_{b\to 0^-}\int_{-1}^{b} f(x-1/x)dx$$
$$+ \lim_{c\to 0^+}\int_c^{1} f(x-1/x)dx + \lim_{d\to\infty}\int_1^{d} f(x-1/x)dx$$

and making the change of variables $x=\frac{1}{2}[y-\sqrt{y^2+4}]$, in the first two integrals, and the change of variables $x=\frac{1}{2}[y+\sqrt{y^2+4}]$, in the second two integrals. Since both of these functions of y have continuous first derivatives on the intervals involved, the change of variables is valid. After the changes of variable, we have four improper integrals. The convergence of each of these integrals is established by a corollary of the Dirichlet Test (Advanced Calculus, R. C. Buck, McGraw-Hill, p. 143). Thus it is permissable to rewrite the first and third of these improper integrals as a single integral by adding the integrands, since they have the same limits from $-\infty$ to 0. The result is $\int_{-\infty}^{0} f(y)dy$. Likewise, the other two integrals combine to give $\int_0^\infty f(y)dy$. In this combining, there is a canceling of a term involving $y/\sqrt{y^2+4}$ because it appears once with a plus sign and once with a minus sign. We have shown both the convergence of $\int_{-\infty}^{\infty} f(x-1/x)dx$ and the desired equality.

B-5 If $a=0$ then $d=1$, and if $a=1$ then $d=0$. In either case $bc=0$ and b or c is 0, while the other is arbitrary. There are $2p-1$ distinct solutions to $bc=0$ and thus the case $a=0$ or $a=1$ accounts for a total of $4p-2$ solutions. If $a\neq 0$ or 1, then d is uniquely determined and $bc\equiv ad\not\equiv 0 \pmod p$ implies that for each $b\neq 0$, there is a unique c, since the integers mod p form a field. Hence for each a in this case, there are $p-1$ solutions. The total number of solutions is $4p-2+(p-2)(p-1)=p^2+p$.

B-6 Let $\{K_n\}$ be any sequence of compact sets of rational numbers. For each n, there is a rational $r_n\notin K_n$, with $0\leqq r_n<1/n$. Otherwise, it would be that K_n contained all rationals in $[0, 1/n]$, and hence some irrationals (since K_n is closed). Let $S=\{0, r_1, r_2, \cdots\}$. Then S is compact and not included in any K_n.

THE THIRTIETH WILLIAM LOWELL PUTNAM MATHEMATICAL COMPETITION

December 6, 1969

A-1 The continuity of $f(x, y)$ implies that the range is connected (i.e., if a, b are in the range and $a<c<b$ then c is in the range). If the range is bounded above and below, then the polynomial $f(x, kx)$ is a constant for each value of k and thus $f(x, y)$ is the constant $f(0, 0)$. Thus the only possibilities are: (i) a single point; (ii) a semi-infinite interval with end-point; (iii) a semi-infinite interval without end-point; and (iv) all real numbers.

Examples are easily given for (i), (ii) and (iv). An example for (iii) is harder to find. One way is to have each cross-section of the surface (for fixed y) be a parabola with a minimum which decreases asymptotically toward some constant as y approaches $\pm\infty$. A suitable example is $(xy-1)^2+x^2$.

A-2 Subtract the first column from every other column. Then add the first row to every other row. The last row now has all zeros except for $(n-1)$ in the first column. D_n is $(-1)^{n-1}(n-1)$ times the minor formed by deleting the first column and last row from the transformed determinant. This minor has only zeros below the main diagonal and thus is equal to the product of its diagonal elements. Hence the minor has value 2^{n-2} and $D_n=(-1)^{n-1}(n-1)2^{n-2}$.

Alternate Solution: From the bottom row of D_{n+1}, subtract $1/(n-1)$ times the first row and $n/(n-1)$ times the nth row. This shows that $D_{n+1} = -[2n/(n-1)]D_n$, for $n>1$. The result follows easily by iteration and the observation that $D_2=-1$.

A-3 Let t be the number of triangles. The sum of all the angles is πt (since it is π for each triangle) and it is also $2\pi m+(n-2)\pi$.

Alternate Solution: Let t be the number of triangles. In Euler's formula $V-E+F=2$, $F=t+1$, and $V=n+m$. Since every edge is on two faces, $2E=3t+n$. Substitution leads directly to the answer $t=2m+n-2$.

Comment: It should have been stated in the problem that the *interior* of the polygon is triangulated. If any of the additional line segments are outside of the polygon, the answer is different.

A-4 A reasonable way to get a series (other than using Riemann sums, which apparently doesn't work) is to write the integrand as a power of e and use the series expansion for e. Then uniform convergence can be applied to interchange integration and summation, and show that

$$\int_0^1 x^x dx = \sum_{m=0}^{\infty} \frac{1}{m!} \int_0^1 x^m (\log x)^m dx.$$

Let $F(m, k) = \int_0^1 x^m (\log x)^k dx$. Integration by parts shows, if applied to a typical term in the summation, why we are interested in $F(m, k)$ and also shows that $F(m, k) = -k/(m+1)F(m, k-1)$ for $m \geqq 0$ and $k \geqq 1$. As a result, $F(m, m) = (-1)^m m!(m+1)^{-m} F(m, 0) = (-1)^m m!(m+1)^{-m-1}$. To get the given formula in the problem, replace $m+1$ by n and adjust the limits on the summation accordingly.

A-5 Subtracting the two equations eliminates $u(t)$ and provides the simpler equation $d(x-y)/dt = 2(x-y)$, which has the solution $x-y = (x_0-y_0)e^{2t}$. If $x_0 \neq y_0$, the right hand side is never zero and so $x=y=0$ can never occur.

For the second part, $x_0 = y_0$. In this case, $x(t) = y(t)$ and so every solution is a parametrization of the line $x=y$. We can attempt to get a solution of the form $x = x_0 - at$, $y = y_0 - at$. This will be a solution if $u(t) = 2(x_0 - at) - a$. By taking $a = x_0/t_0$, $x=y=0$ at $t=t_0$.

A-6 Let $\bar{y} = \lim_{n\to\infty} y_n$ and set $\bar{x} = \bar{y}/3$. We will show that $\bar{x} = \lim_{n\to\infty} x_n$. For any $\epsilon > 0$ there is an N such that for all $n > N$, $|y_n - \bar{y}| < \epsilon/2$.

$$\epsilon/2 > |y_n - \bar{y}| = |x_{n-1} + 2x_n - 3\bar{x}| = |2(x_n - \bar{x}) + (x_{n-1} - \bar{x})|$$
$$\geqq 2|x_n - \bar{x}| - |x_{n-1} - \bar{x}|.$$

This may be rewritten as $|x_n - \bar{x}| < \epsilon/4 + \frac{1}{2}|x_{n-1} - \bar{x}|$, which can be iterated to give

$$|x_{n+m} - \bar{x}| < \epsilon/4\left(\sum_{i=0}^{m} 2^{-i}\right) + 2^{-(m+1)}|x_{n-1} - \bar{x}| < \epsilon/2 + 2^{-(m+1)}|x_{n-1} - \bar{x}|.$$

By taking m large enough, $2^{-(m+1)}|x_{n-1} - \bar{x}| < \epsilon/2$. Thus for all sufficiently large k, $|x_k - \bar{x}| < \epsilon$.

B-1 The condition $24 | n+1$ is equivalent to $n \equiv -1 \pmod 3$ and $n \equiv -1 \pmod 8$. Let d be a divisor of n, then $d \equiv 1$ or $2 \pmod 3$ and $d \equiv 1, 3, 5$ or $7 \pmod 8$. Since $d(n/d) = n \equiv -1 \pmod 3$ or $\pmod 8$, the only possibilities are:

$d \equiv 1,$	$n/d \equiv 2 \pmod 3$	or vice versa
$d \equiv 1,$	$n/d \equiv 7 \pmod 8$	or vice versa
$d \equiv 3,$	$n/d \equiv 5 \pmod 8$	or vice versa.

In every case, $d + n/d \equiv 0 \pmod 3$ and $\pmod 8$. Thus $d + n/d$ is a multiple of 24. Note that $d \neq n/d$ and thus no divisor is used twice in the pairing, so the sum of all the divisors is a multiple of 24.

B-2 The number of elements in a subgroup is a divisor of the order of the group. Thus a proper subgroup can have no more than half of all the elements. Two subgroups always have the identity in common and hence their union cannot be the entire group.

An example for the second part of the problem is the Klein group, which has an identity and three elements x, y, z of order two. The product of any two distinct elements from $\{x, y, z\}$ is the third. This group is the union of three proper subgroups.

Alternate Solution: Let $G=H\cup K$, with H and K proper subgroups. There exists $k\in K$ with $k\notin H$. None of the elements in kH are in H and so $kH\subset K$. Hence $H\subset k^{-1}K=K$ and $K=H\cup K=G$, a contradiction.

B-3 The first relation implies that

$$T_n = \frac{(n-1)(n-3)\cdots 3}{(n-2)(n-4)\cdots 2}\cdot\frac{1}{T_1} \qquad \text{if } n \text{ is even,}$$

$$T_n = \frac{(n-1)(n-3)\cdots 2}{(n-2)(n-4)\cdots 1}\cdot T_1 \qquad \text{if } n \text{ is odd.}$$

If n is odd,

$$\frac{T_n}{T_{n+1}} = (T_1)^2\cdot\frac{2}{1}\,\frac{2}{3}\,\frac{4}{3}\,\frac{4}{5}\,\frac{6}{5}\cdots\frac{(n-1)}{(n-2)}\,\frac{(n-1)}{n}.$$

The Wallis product is $\pi/2=\frac{2}{1}\frac{2}{3}\frac{4}{3}\frac{4}{5}\frac{6}{5}\cdots$. After an even number of factors the partial product is less than $\pi/2$ and after an odd number of factors the partial product is greater than $\pi/2$. Thus for the case when n is odd, $T_n/T_{n+1}<\frac{1}{2}\pi T_1^2$. A similar calculation shows that, when n is even, $T_n/T_{n+1}<2/\pi T_1^2$. Since the limit of $T_n/T_{n+1}=1$, 1 is less than or equal to both $\frac{1}{2}\pi T_1^2$ and its reciprocal. This implies that $\pi T_1^2=2$.

B-4 Place the curve so that its endpoints lie on the x-axis. Then take the smallest rectangle with sides parallel to the axes which covers the curve. Let its horizontal and vertical dimensions be a and b respectively. Let P_0 and P_5 be the endpoints of the curve, and let P_1, P_2, P_3, and P_4 be the points on the curve, in the order named, which lie one on each of the four sides of the rectangle. Draw the broken line $P_0P_1P_2P_3P_4P_5$. This line has length at most one. The horizontal components of the segments of this broken line add up at least to a, since one of the vertices of the broken line lies on the left end of the rectangle and one on the right end. The vertical segments add to at least $2b$ since we start and finish on the x-axis and go to both the top and bottom sides. This implies that the total length of the broken line is at least $(a^2+4b^2)^{1/2}$.

We now have that a and b both lie between 0 and 1 and that $a^2+4b^2\leqq 1$. Under these conditions the product ab is a maximum for $a=\frac{1}{2}\sqrt{2}$, $b=\frac{1}{4}\sqrt{2}$ and so the maximum of ab is $\frac{1}{4}$. Thus the area of the rectangle we have constructed is at most $\frac{1}{4}$.

B-5 The following proof shows it is not necessary to stipulate that the a_n be integers. Suppose for some $\epsilon > 0$ there are $x_j \to \infty$ with $k(x_j)/x_j \geqq \epsilon$. Note that if $1 \leqq n \leqq k(x_j)$, then (because the a_n increase) $a_n \leqq a_{k(x_j)} \leqq x_j$ and $1/a_n \geqq 1/x_j$. Now for any positive integer N,

$$\sum_{n=N}^{\infty} 1/a_n \geqq \sup_j \sum_{n=N}^{k(x_j)} 1/a_n \geqq \sup_j \frac{k(x_j) - N}{x_j} \geqq \sup_j (\epsilon - N/x_j) = \epsilon.$$

But this contradicts the convergence of $\sum_{n=1}^{\infty} 1/a_n$, which implies

$$\lim_{N \to \infty} \sum_{n=N}^{\infty} 1/a_n = 0.$$

B-6 Observe that $ABAB = 9AB$. AB is of rank two so A is onto and B is one-to-one. Hence there exist matrices A' and B' such that $A'A = I = BB'$, where I is the 2×2 identity matrix. Then $A'(ABAB)B' = BA = 9I$.

Alternate Solution: $(AB)^2 = 9AB$. The rank of BA is greater than or equal to the rank of $A(BA)B$, which is 2. Thus BA is nonsingular. But $(BA)^3 = B(AB)^2A = B(9AB)A = 9(BA)^2$ and the result follows since BA has an inverse.

THE THIRTY-FIRST WILLIAM LOWELL PUTNAM MATHEMATICAL COMPETITION

December 5, 1970

A-1 Note that $e^{ax} \cos bx$ is the real part of $e^{(a+ib)x}$. Thus the power series is

$$e^{ax} \cos bx = \sum_{n=0}^{\infty} \mathrm{Re}\{(a+ib)^n\}\frac{x^n}{n!} \cdot$$

In this form, it is easily seen that if x^n has a zero coefficient, then x^{kn} has a zero coefficient for every odd value of k.

A-2 Let $(x, y) = (r \cos \theta, r \sin \theta)$, $r > 0$, be a point of the locus. Then

$$r = \frac{|A \cos^2 \theta + B \sin \theta \cos \theta + C \sin^2 \theta|}{|D \cos^3 \theta + E \cos^2 \theta \sin \theta + F \cos \theta \sin^2 \theta + G \sin^3 \theta|} \cdot \tag{1}$$

The denominator of (1) is less than or equal to $|D| + |E| + |F| + |G|$, whereas the numerator has a positive minimum

$$N = \frac{|A+C| - \sqrt{(A-C)^2 + B^2}}{2},$$

since $B^2 < 4AC$. Therefore

$$r \geqq \frac{N}{|D| + |E| + |F| + |G|} = \delta$$

and there are no points of the locus within $0 < r < \delta$.

Alternate Solution: Set $H(x, y)$ equal to the polynomial on the left hand side of the given equation. The standard theory for maxima or minima of functions of two variables can be used together with the condition $B^2 < 4AC$ to show that $H(x, y)$ has a local maximum or a local minimum at $(0, 0)$.

A-3 If x is an integer then $x^2 \equiv 0, 1, 4, 6$ or $9 \pmod{10}$. The case $x^2 \equiv 0 \pmod{10}$ is eliminated by the statement of the problem. If $x^2 \equiv 11, 55$ or $99 \pmod{100}$, then $x^2 \equiv 3 \pmod 4$ which is impossible. Similarly, $x^2 \equiv 66 \pmod{100}$ implies $x^2 \equiv 2 \pmod 4$ which is also impossible. Therefore $x^2 \equiv 44 \pmod{100}$. If $x^2 \equiv 4444 \pmod{10{,}000}$, then $x^2 \equiv 12 \pmod{16}$, but a simple check shows that this is impossible. Finally note that $(38)^2=1444$.

A-4 For $\epsilon>0$, let N be sufficiently large so that $|x_n-x_{n-2}|<\epsilon$ for all $n\geqq N$. Note that for any $n>N$,

$$x_n - x_{n-1} = (x_n - x_{n-2}) - (x_{n-1} - x_{n-3}) + (x_{n-2} - x_{n-3}) - \cdots$$
$$\pm (x_{N+1} - x_{N-1}) \mp (x_N - x_{N-1}).$$

Thus $|x_n-x_{n-1}| \leqq (n-N)\epsilon+|x_N-x_{N-1}|$ and $\lim_{n\to\infty} (x_n-x_{n-1})/n=0$.

A-5 Since parallel cross sections of the ellipsoid are always similar ellipses, any circular cross section can be increased in size by taking a parallel cutting plane passing through the center. Every plane through (0, 0, 0) which makes a circular cross section must intersect the y-z plane. But this means that a diameter of the circular cross section must be a diameter of the ellipse $x=0$, $y^2/b^2+z^2/c^2=1$. Hence the radius of the circle is at most b. Similar reasoning with the x-y plane shows that the radius of the circle is at least b, so that any circular cross section formed by a plane through (0, 0, 0) must have radius b, and this will be the required maximum radius. To show that circular cross sections of radius b actually exist, consider all planes through the y-axis. It can be verified that the two planes given by $a^2(b^2-c^2)z^2=c^2(a^2-b^2)x^2$ give circular cross sections of radius b.

A-6 Let x be selected from $[0, L_1]$, y from $[0, L_2]$, z from $[0, L_3]$, and assume $L_3\geqq L_2\geqq L_1$. Let $X=\min(x, y, z)$.

$$L_1L_2L_3E[X] = \int_0^{L_1}\int_0^{L_2}\int_0^{L_3} X\,dz\,dy\,dx$$

$$= \int_0^{L_1}\int_0^{L_2}\left\{\int_0^{\mu} z\,dz + \int_{\mu}^{L_3} \mu\,dz\right\} dy\,dx, \quad \text{where } \mu = \min(x, y),$$

$$= \int_0^{L_1}\int_0^{L_2}\{L_3\mu - \tfrac{1}{2}\mu^2\}\,dy\,dx$$

$$= \int_0^{L_1}\left\{\int_0^{x}(L_3y - \tfrac{1}{2}y^2)dy + \int_x^{L_2}(L_3x - \tfrac{1}{2}x^2)dy\right\} dx$$

$$= \cdots = \frac{1}{2}L_1^2L_2L_3 - \frac{1}{6}L_1^3(L_2 + L_3) + \frac{1}{12}L_1^4.$$

Alternate Solution: For $0\leqq a\leqq L$,

$$P(X \leqq a) = P(x \leqq a) + P(y \leqq a) + P(z \leqq a) - P(x \leqq a)P(y \leqq a)$$
$$- P(x \leqq a)P(z \leqq a) - P(y \leqq a)P(z \leqq a)$$
$$+ P(x \leqq a)P(y \leqq a)P(z \leqq a)$$
$$= \frac{a}{L_1} + \frac{a}{L_2} + \frac{a}{L_3} - \left(\frac{a^2}{L_1L_2} + \frac{a^2}{L_2L_3} + \frac{a^2}{L_3L_1}\right) + \frac{a^3}{L_1L_2L_3}\cdot$$

The answer follows easily from the formula

$$E[X] = \int_0^{L_1} a \frac{dP(X \leq a)}{da} da.$$

B-1 Let

$$a_n = \frac{1}{n^4} \prod_{i=1}^{2n} (n^2 + i^2)^{1/n}.$$

Then

$$\log a_n = \frac{1}{n} \sum_{i=1}^{2n} \log\left(1 + \frac{i^2}{n^2}\right),$$

and

$$\operatorname*{limit}_{n\to\infty} \log a_n = \int_0^2 \log(1 + x^2)dx = 2 \log 5 - 4 + 2 \arctan 2.$$

B-2 Let $P(t) = at^3 + bt^2 + ct + d$. The equation

$$\frac{1}{2T} \int_{-T}^{T} P(t)dt = \frac{1}{2} \{P(t_1) + P(t_2)\}$$

is satisfied for all values of a, b, c, and d if and only if $t_2 = -t_1 = \pm T/\sqrt{3}$. If $T = 3$ hrs, $T/\sqrt{3} \approx 1$ hr, 43.92 min. Therefore, in the case considered, the critical times are 1 hour 44 minutes each side of noon.

B-3 Let $y_n \to y$ with $(x_n, y_n) \in S$ for all n. The Bolzano-Weierstrass Theorem implies that a subsequence $x_{k(n)} \to x$. Then $y_{k(n)} \to y$ and since S is closed, $(x, y) \in S$. Thus y is in the projection of S on the y-axis.

B-4 Converting units to feet and seconds, we have $0 \leqq v(t) \leqq 132$ for all $t \in [0, 60]$. Suppose $|v'(t)| < 6.6$ for all $t \in [0, 60]$. Then $v(t) = \int_0^t v' < 6.6t$, and $v(t) = \int_t^{60} -v' < 6.6(60 - t)$ for all $t \in [0, 60]$. Thus

$$5280 = \int_0^{60} v(t)dt < \int_0^{60} \min\{6.6t, 6.6(60 - t), 132\} dt.$$

This last integral is the area under a trapezoid and equals the value 5280, which is a contradiction.

B-5 Clearly u_n is continuous. So, if F is continuous, then $u_n \circ F$ is the composition of continuous functions and hence is continuous. Conversely, suppose $u_n \circ F$ is continuous for all n. To prove F is continuous it is enough to show $F^{-1}[(a, b)]$ is open for every bounded interval (a, b). Let $n > \max(|a|, |b|)$. Then $u_n^{-1}[(a, b)] = (a, b)$ so

$$F^{-1}[(a, b)] = F^{-1}[u_n^{-1}\{(a, b)\}] = (u_n \circ F)^{-1}[(a, b)],$$

which is an open set by the continuity of $u_n \circ F$.

B-6 Since the quadrilateral is circumscribable, $a+c=b+d$. Let k be the length of a diagonal and angles α and β selected so that $k^2=a^2+b^2-2ab\cos\alpha=c^2+d^2-2cd\cos\beta$. If we subtract $(a-b)^2=(c-d)^2$, we obtain

$$(1) \qquad 2ab(1-\cos\alpha)=2cd(1-\cos\beta).$$

From $A=\frac{1}{2}ab\sin\alpha+\frac{1}{2}cd\sin\beta=\sqrt{abcd}$,

$$4A^2=4abcd=a^2b^2(1-\cos^2\alpha)+c^2d^2(1-\cos^2\beta)+2abcd\sin\alpha\sin\beta.$$

Using (1) twice on the right hand side,

$$4abcd=ab(1+\cos\alpha)cd(1-\cos\beta)+cd(1+\cos\beta)ab(1-\cos\alpha)+2abcd\sin\alpha\sin\beta.$$

On simplifying, $4=2-2\cos(\alpha+\beta)$, which implies that $\alpha+\beta=\pi$ and so the quadrilateral is cyclic.

THE THIRTY-SECOND WILLIAM LOWELL PUTNAM MATHEMATICAL COMPETITION

December 4, 1971

A–1 The set of all lattice points can be divided into eight classes according to the parities of the coordinates, namely, (odd, odd, odd), (odd, odd, even), etc. With nine lattice points some two, say P and Q, belong to the same class. The midpoint of the segment PQ is a lattice point.

A–2 $P(0) = 0$, $P(1) = [P(0)]^2 + 1 = 1$, $P(2) = [P(1)]^2 + 1 = 2$, $P(5) = [P(2)]^2 + 1 = 5$, $P(5^2 + 1) = [P(5)]^2 + 1 = 26$, etc. Thus the polynomial $P(x)$ agrees with x for more values than the degree of $P(x)$, so $P(x) \equiv x$.

A–3 For a triangle with sides a, b, c, area $= A$ and circumradius $= R$ we have $abc = 4\,RA$. But if the vertices are lattice points the determinant formula (or Pick's Theorem or direct calculation) for the area shows that $2A$ is an integer. Hence $2A \geqq 1$, so that $abc \geqq 2R$. To obtain the formula $abc = 4RA$ note that if α is the angle opposite side a, then side a subtends an angle 2α at the center and $a = 2R \sin\alpha$, $A = \frac{1}{2} bc \sin\alpha$.

A–4 In the expansion of $(x + y)^n(x^2 - (2 - \varepsilon)xy + y^2)$ the coefficient of $x^{k+1}y^{n+1-k}$ is

$$\binom{n}{k-1} - (2 - \varepsilon)\binom{n}{k} + \binom{n}{k+1}$$

$$= \binom{n}{k}\left\{\frac{k}{n-k+1} + \frac{n-k}{k+1} - (2-\varepsilon)\right\}.$$

Now for fixed n consider the expression

$$\phi(k) = \frac{k}{n-k+1} + \frac{n-k}{k+1} - (2-\varepsilon).$$

If k is taken to be a continuous positive variable

$$\phi'(k) = \frac{(n+1)\{(k+1)^2 - (n-k+1)^2\}}{(n-k+1)^2(k+1)^2}.$$

Hence $\phi'(k) = 0$ at $k = n/2$ and it follows easily that $\phi(k)$ is minimum at $k = n/2$.

We needn't consider end point minima since it easily follows that for $n > 2$ the polynomial has its first two and last two coefficients positive. We may also note that if the two mid-terms in the expansion are non-positive for a given odd value of n then for the next larger value of n the mid-term remains non-positive. Hence if the mid-coefficients become positive, the first value of n for which this occurs is odd. Now if n is odd and $k = \frac{1}{2}(n+1)$ then $\phi(k) = \dfrac{n-1}{n+3} - 1 + \varepsilon$, and $\phi(k) > 0$ for $n > \dfrac{4}{\varepsilon} - 3$. If $\varepsilon = .002$, $n > 1997$ and n is odd. Hence the minimum n for which all terms are positive is 1999.

A-5 The attainable scores are those non-negative integers expressible in the form $xa + yb$ with x and y non-negative integers. If a and b are not relatively prime there are infinitely many non-attainable scores. Hence $(a, b) = 1$. It will be shown that the number of non-attainable scores is $\frac{1}{2}(a-1)(b-1)$.

If m is an attainable score, the line $ax + by = m$ passes through at least one lattice point in the closed first quadrant. Because a and b are relatively prime, the lattice points on a line $ax + by = m$ are at a horizontal distance of b. The first-quadrant segment of $ax + by = m$ has a horizontal projection of m/a and thus every score $m \geqq ab$ is attainable. Every non-attainable score must satisfy $0 \leqq m < ab$.

If $0 \leqq m < ab$, the first-quadrant segment of the line $ax + by = m$ has a horizontal projection less than b, and so contains at most one lattice point. Thus there is a one-to-one correspondence between lattice points (x, y) with $0 \leqq ax + by < ab$ in the first quadrant and attainable scores with $0 \leqq m < ab$. The closed rectangle $0 \leqq x \leqq b$, $0 \leqq y \leqq a$ contains $(a+1)(b+1)$ lattice points, so the number of lattice points in the first quadrant with $0 \leqq ax + by < ab$ is $\frac{1}{2}(a+1)(b+1) - 1$. This is the number of attainable scores with $0 \leqq m < ab$. Hence the number of non-attainable scores in this range (which is all of them) is $ab - \frac{1}{2}(a+1)(b+1) + 1 = \frac{1}{2}(a-1)(b-1)$.

In our given example $70 = (a-1)(b-1) = 1(70) = 2(35) = 5(14) = 7(10)$. The conditions $a > b$, $(a, b) = 1$ yield two possibilities $a = 71$, $b = 2$ and $a = 11$, $b = 8$. Since $58 = 71(0) + 2(29)$, the first of these alternatives is eliminated. The line $11x + 8y = 58$ passes through $(6, -1)$ and $(-2, 10)$ and thus does not pass through a lattice point in the first quadrant. The unique solution is $a = 11$, $b = 8$.

A-6 The case $n = 2$ shows that c is non-negative. If the ordinary mean value theorem is applied to x^c on the interval $[u, u+1]$ there is a ξ with $u < \xi < u+1$ such that $c\,\xi^{c-1} = (u+1)^c - u^c$. For any positive integer u the right hand side is a positive integer. Now, in the case $0 < c < 1$, u could be taken large enough so $u^{c-1} < 1/c$ and so $c\,\xi^{c-1} < 1$. Thus the mean value theorem for the first derivative eliminates all c with $0 < c < 1$.

There is an extension of the mean value theorem which states that if $f(x)$ is k-times differentiable in $[a, b]$ then there is a ξ, $a < \xi < b$, such that $h^k f^{(k)}(\xi) = \Delta^k f(a)$, where $h = \dfrac{b-a}{k}$ and Δ^k is the k-th difference for intervals spaced h apart. Take k as the unique integer such that $k-1 \leqq c < k$ and apply this extension of the mean value theorem on the interval $[u, u+k]$. There is a ξ with $u < \xi < u+k$ such that

$$c(c-1)(c-2)\cdots(c-k+1)\,\xi^{c-k} = \Delta^k f(u).$$

The right hand side is an integer, and by taking u sufficiently large ξ^{c-1} becomes sufficiently small so that the left hand side, though non-negative, is less than 1. Hence $c(c-1)(c-2)\cdots(c-k+1) = 0$ and so $c = k-1$.

B–1 Using the given laws we have

$$x \circ y = (x \circ y) \circ (x \circ y) = [(x \circ y) \circ x] \circ y = [(y \circ x) \circ x] \circ y$$
$$= [(x \circ x) \circ y] \circ y = (x \circ y) \circ y = (y \circ y) \circ x = y \circ x.$$

From this commutative law we obtain

$$(x \circ y) \circ z = (y \circ z) \circ x = x \circ (y \circ z).$$

B–2 In the given functional equation

$$F(x) + F\left(\frac{x-1}{x}\right) = 1 + x \tag{1}$$

we substitute $\dfrac{x-1}{x}$ for x, obtaining

$$F\left(\frac{x-1}{x}\right) + F\left(\frac{-1}{x-1}\right) = \frac{2x-1}{x}. \tag{2}$$

Also in (1), we substitute $\dfrac{-1}{x-1}$ for x and obtain

$$F\left(\frac{-1}{x-1}\right) + F(x) = \frac{x-2}{x-1}. \tag{3}$$

Adding (1) and (3) and subtracting (2) gives

$$2F(x) = 1 + x + \frac{x-2}{x-1} - \frac{2x-1}{x} = \frac{x^3 - x^2 - 1}{x(x-1)}.$$

$$F(x) = \frac{x^3 - x^2 - 1}{2x(x-1)}. \tag{4}$$

That $F(x)$, defined in (4), does satisfy the given functional equation is easily verified. Therefore (4) is the only solution of the problem.

B–3 At time t, car 1 has completed $[t]$ laps and car 2 has completed $[t-T]$ laps. The problem is to find values of $t \geqq T$ for which $[t] = 2[t-T]$.

Let $T = k + \delta$, where $0 \leqq \delta < 1$, k an integer. Consider any integral interval

$[m, m+1]$ and let $m \leqq t < m+1$. Then $t = m + \varepsilon$, where $0 \leqq \varepsilon < 1$. Then the equation to be solved becomes

$$[t] = m = 2[t - T] = 2[m + \varepsilon - (k + \delta)] = 2[m - k + \varepsilon - \delta].$$

Thus $m = 2(m - k)$, if $\varepsilon \geqq \delta$ and $m = 2(m - k - 1)$, if $\varepsilon < \delta$. If $1 > \varepsilon \geqq \delta$, then $m = 2k$ and the equation is satisfied during $[2k + \delta, 2k + 1]$, which has length $1 - \delta$.

If $0 \leqq \varepsilon < \delta$, then $m = 2k + 2$ and the equation is satisfied during $[2k + 2, 2k + 2 + \delta]$ which has length δ. Therefore the total length is $1 - \delta + \delta = 1$.

Comment: The problem should have been more explicit by stating "after the start of the second car" instead of "during the motion". The solution is given for this interpretation, whereas, if $t < T$, $[t - T]$ is negative but the second car would have completed zero laps.

B–4 We take the radius of the sphere as unity and denote the constant sum $\widehat{PA} + \widehat{PB}$ by $2a$. To avoid trivial and degenerate cases we assume that $0 < \widehat{AB} < \pi$ and that $\widehat{AB} < 2a < 2\pi - \widehat{AB}$.

The case $2a > \pi$ can be reduced to the case $2a < \pi$. For, if A' and B' are the points diametrically opposite to A and B then $\widehat{PA} + \widehat{PB} = 2a$ if and only if $\widehat{PA'} + \widehat{PB'} = 2\pi - 2a$; that is, the spherical ellipses $\widehat{PA} + \widehat{PB} = 2a$ and $\widehat{PA'} + \widehat{PB'} = 2\pi - 2a$ are identical. Since $\min(2a, 2\pi - 2a) \leqq \pi$, we may assume without loss of generality that $2a \leqq \pi$.

Let A and B lie on the equator. There are two points V_1 and V_2 (the "vertices") on the equator which lie on the spherical ellipse. Obviously, $\widehat{V_1V_2} = 2a$. The "center" of the spherical ellipse (common midpoint of the arcs $\widehat{AB}$ and $\widehat{V_1V_2}$) will be denoted by C.

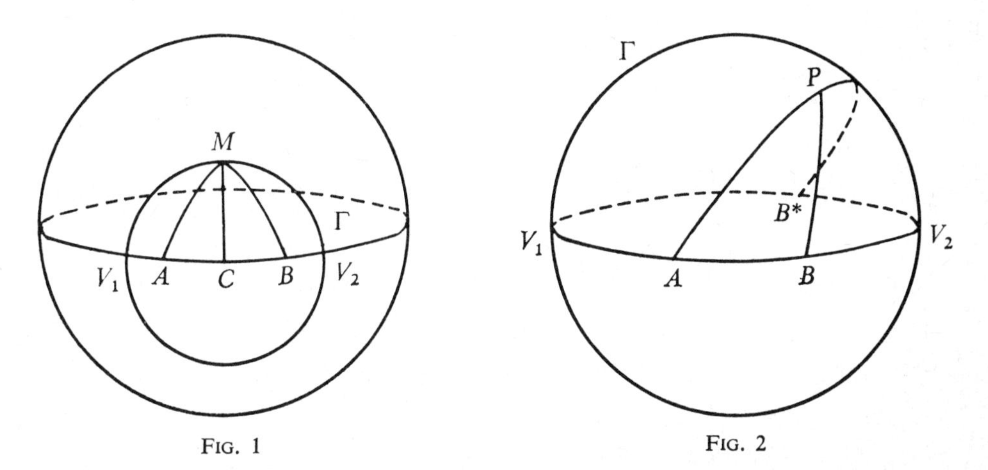

FIG. 1 FIG. 2

We first treat the case $2a < \pi$ and show that in this case the spherical ellipse cannot be a circle. Assume it were a circle; call it Γ (see Figure 1). Γ would have to be symmetric with respect to the equatorial plane, thus lie in a plane perpendicular to the equatorial plane. Γ would also have to pass through the vertices. Therefore its spherical diameter would be $\widehat{V_1V_2} = 2a$ and its spherical radius would be equal to a. The spherical center of Γ would be C, the center of the ellipse. Let M be one of the two points on Γ which lie half-way between the two vertices. Then, since M is supposed to be a point on the spherical ellipse, $2a = \widehat{MA} + \widehat{MB} > 2\widehat{MC} = 2a$ (note that MAC is a right spherical triangle with the right angle at C and with side $\widehat{MC} = a < \frac{1}{2}\pi$). Contradiction shows that the only possible spherical ellipses which are circles must occur when $2a = \pi$.

In case $2a = \pi$, V_1 and V_2 are diametrically opposite points on the equator. We shall show that the great circle Γ through the vertices and perpendicular to the equatorial plane is identical with the spherical ellipse $\widehat{PA} + \widehat{PB} = \pi$. To see this, let B^* be the reflection of B about the plane of Γ. B^* is on the equator diametrically opposite to A (see Fig. 2). Let P be an arbitrary point on the sphere, and draw the great circle through A, P and B^*. Then $\widehat{PA} + \widehat{PB^*} = \pi$. Hence, $\widehat{PA} + \widehat{PB} = \pi$ if and only if $\widehat{PB} = \widehat{PB^*}$, that is, if and only if P is on Γ. This shows that Γ is the spherical ellipse $\widehat{PA} + \widehat{PB} = \pi$, as stated above.

Thus the only circles on the sphere that are spherical ellipses are the great circles. For any given great circle Γ the foci can be any two points A and B which lie on the same great circle perpendicular to Γ, on the same side of Γ and at equal distances from Γ. The equation of any such spherical ellipse is $\widehat{PA} + \widehat{PB} = \pi$.

B–5 We put $z = x + iy$. Then both differential equations can be combined into one, namely

$$z'' - iz' + 6z = 0. \tag{1}$$

This is a standard linear equation of the second order with constant coefficients and has the general solution

$$z(t) = c_1 e^{3it} + c_2 e^{-2it}.$$

The initial conditions imply $z'(0) = 0$ or $3ic_1 - 2ic_2 = 0$. We may set $c_1 = 2A$ and $c_2 = 3A$, where A is any complex number. The general solution of the given system is

$$z(t) = 2Ae^{3it} + 3Ae^{-2it}. \tag{2}$$

If $A = R\,e^{i\alpha}$, then a rotation of axes through the angle α produces

$$Z(t) = 2R\,e^{3it} + 3R\,e^{-2it} \tag{3}$$

or in rectangular form

$$\begin{aligned} X(t) &= 2R\cos(3t) + 3R\cos(2t) \\ Y(t) &= 2R\sin(3t) - 3R\sin(2t). \end{aligned} \tag{4}$$

This is the standard form for a hypocycloid when the radius of the rolling circle is $3R$ and the fixed circle is of radius $5R$. On time reversal it becomes the standard equations of a hypocycloid with radius of the rolling circle of $2R$ and the radius of the fixed circle of $5R$.

B–6 Set

$$S(x) = \sum_{n=1}^{x} \frac{\delta(n)}{n}.$$

Note that $\delta(2m+1) = 2m+1$, $\delta(2m) = \delta(m)$ and that $S(2x+1) = S(2x)+1$. Dividing the summation for $S(2x)$ into even and odd values of the index produces the following relation:

$$S(2x) = \sum_{m=1}^{x} \frac{\delta(2m)}{2m} + \sum_{m=1}^{x} \frac{\delta(2m-1)}{2m-1} = \tfrac{1}{2}S(x) + x.$$

If we denote $S(x) - \dfrac{2x}{3}$ by $F(x)$, the above relations translate into

$$F(2x) = \tfrac{1}{2}F(x), \text{ and } F(2x+1) = F(2x) + \tfrac{1}{3}.$$

Now induction can be used to show that $0 < F(x) < \frac{2}{3}$, for all positive integers x. This result is sharper than that requested.

THE THIRTY-THIRD WILLIAM LOWELL PUTNAM MATHEMATICAL COMPETITION

December 2, 1972

A–1 For a given n and r, in order for the first three binomial coefficients to be in arithmetic progression, we must have

$$2\binom{n}{r+1} = \binom{n}{r} + \binom{n}{r+2} \tag{1}$$

or equivalently

$$2 = \frac{r+1}{n-r} + \frac{n-r-1}{r+2}. \tag{2}$$

The condition that the last three given binomial coefficients are in arithmetic progression is found from (1) by replacing r by $r+1$. Consequently both r and $r+1$ must satisfy equation (2) if all four terms are in arithmetic progression.

Note that the two terms in equation (2) are interchanged if r is replaced by $n-r-2$. Thus the quadratic equation (2) has roots

$$r, r+1; n-r-3, n-r-2.$$

Since (2) can have only two roots, $r = n-r-3$ and $n = 2r+3$. The four binomial coefficients must be

$$\binom{2r+3}{r}, \binom{2r+3}{r+1}, \binom{2r+3}{r+2}, \binom{2r+3}{r+3}$$

which are the four middle terms. They cannot be in arithmetic progression since binomial coefficients increase to the middle term(s) and then decrease.

A–2 Label the given laws (1) and (2), respectively.

I. We first show that

$$(x * y) * x = y. \tag{3}$$

This follows from $(x * y) * x = (x * y)\,[(x * y) * y] = y$. (First apply (2) with x and y interchanged; then apply (1) with x replaced by $x * y$.)

We now obtain

(4) $$y * x = [(x * y) * x] * x = x * y.$$

(First apply (3); then apply (2) with y replaced by $x * y$.) This proves that $*$ is commutative.

II. Let S be the set of all integers. Define $x * y = -x - y$. Then

(5) $$x * (y * z) = -x + y + z; \quad (x * y) * z = x + y - z.$$

It follows from (5) that, in the first place, (1) and (2) hold and, secondly, $*$ fails to be associative: simply choose $x \neq z$ in (5).

Alternate Solution, Part I (suggested by Martin Davis):

Write the equation $x * y = z$ as $P(x, y, z)$. Then law (1) may be written "If $x * y = z$ then $x * z = y$" or

(6) $$P(x, y, z) \text{ implies } P(x, y, z).$$

Similarly, the law (2) may be written

(7) $$P(y, x, z) \text{ implies } P(z, x, y).$$

These two implications, (6) and (7), show that the permutations (23) and (13) on the location of the variables in $P(x, y, z)$ are permitted. Since (13), (23) generate the symmetric group S_3, we find (12) is also permitted.

Thus, $P(x, y, z)$ implies $P(y, x, z)$ or $x * y = z$ implies $y * x = z$, which means $x * y = y * x$.

A–3 A function is "supercontinuous" if and only if it is affine, $f(x) = Ax + B$. The sufficiency is trivial (and was worth 1 point in the grading). For the necessity: First we note that it is *not* assumed that $f(C\text{-limit}) = C\text{-limit } (f)$ (otherwise the solution could be materially simplified). The essential steps are to show, that if f is supercontinuous, then (1) f is continuous, and (2) $f((a + b)/2) = (f(a) + f(b))/2$ for all a, b. These two statements imply that f is affine. The proofs of (1) and (2) are similar; we give (2) (which is the harder). Set $c = (a + b)/2$, and suppose $f(c) \neq (f(a) + f(b))/2$. Imagine any sequence of integers N_i which "grows very rapidly"; say let N_{i+1} exceed $2^i N^i$. Then construct a sequence of points $\{x_n\}$ as follows: Break the sequence into blocks, alternating between

$$\{x_n\} = a, b, a, b, a, b, \cdots$$

and

$$\{x_n\} = c, c, c, c, c, c, \cdots,$$

the ab pattern holding for $N_{2i-1} \leqq n < N_{2i}$, and the c pattern holding for $N_{2i} \leqq n < N_{2i+1}$. Then $\{x_n\}$ has the C limit c, but the averages of $\{f(x_n)\}$ oscillate (because the lengths of the blocks $N_i \leqq n < N_{i+1}$ increase very fast, and $f(c) \neq$ the average of $f(a)$ and $f(b)$). Thus the C-limit of $\{f(x_n)\}$ does not exist, a contradiction.

A–4 Let the square of sidelength $2R$ have the vertices $(\pm R\sqrt{2}, 0)$ and $(0, \pm R\sqrt{2})$. The ellipse

(1) $$\frac{x^2}{a^2} + \frac{y^2}{b^2} = 1$$

with $0 \leqq b \leqq a \leqq R\sqrt{2}$ has the line $x + y = R\sqrt{2}$ as a tangent if and only if the quadratic equation $x^2/a^2 + (R\sqrt{2} - x)^2/b^2 = 1$ has a double root. It can be verified that its discriminant vanishes if and only if $a^2 + b^2 = 2R^2$. As a varies from R to $R\sqrt{2}$ and b varies from R to 0, the curve (1) varies from the circle of radius R through all the non-circular ellipses inscribed in the square to the degenerate "flat" ellipse lying on the x-axis.

Let $4L$ denote the length of the ellipse $x = a \cos t$, $y = b \sin t$, $0 \leqq t \leqq 2\pi$. Then

$$L = \int_0^{\pi/2} [a^2 \sin^2 t + b^2 \cos^2 t]^{\frac{1}{2}} dt = \int_0^{\pi/2} [\tfrac{1}{2}a^2(1 - \cos 2t) + \tfrac{1}{2}b^2(1 + \cos 2t)]^{\frac{1}{2}} dt$$

$$= \int_0^{\pi/2} [R^2 - \tfrac{1}{2}c^2 \cos 2t]^{\frac{1}{2}} dt,$$

where $c^2 = a^2 - b^2$. The last integral we split into one from 0 to $\pi/4$ and one from $\pi/4$ to $\pi/2$, and in the latter we substitute $t = \pi/2 - t'$, obtaining

(2) $$L = \int_0^{\pi/4} \{[R^2 - \tfrac{1}{2}c^2 \cos 2t]^{\frac{1}{2}} + [R^2 + \tfrac{1}{2}c^2 \cos 2t]^{\frac{1}{2}}\} dt.$$

Note that $\cos 2t > 0$ for $0 \leqq t < \pi/4$.

Now the function $f(u) = (p - u)^{\frac{1}{2}} + (p + u)^{\frac{1}{2}}$ decreases in the interval $0 \leqq u \leqq p$, because $2f'(u) = -(p - u)^{-\frac{1}{2}} + (p + u)^{-\frac{1}{2}} < 0$ for $0 < u < p$. Thus the integral in (2) as a function of c has its largest value when $c = 0$, that is, for the inscribed circle.

To show that an ellipse inscribed in the square must have its axes along the diagonals of the square, we choose the square as having sides $u = \pm R$ and $v = \pm R$ and the ellipse as having the equation

$$Au^2 + Buv + Cv^2 + Du + Ev + F = 0,$$

where

(1) $$4AC - B^2 > 0.$$

Taking the "highest," "lowest," "rightest," and "leftest" points on the ellipse, we see that all four sides of the square must be tangents to the ellipse.

The line $u = R$ is a tangent if and only if the equation $Cv^2 + (BR + E)v + (Ar^2 + Dr + F) = 0$ has a double root or

(2) $$(BR + E)^2 - 4C(AR^2 + DR + F) = 0.$$

The corresponding conditions for $u = -R$, $v = R$ and $v = -R$ are

(3) $$(-BR + E)^2 - 4C(AR^2 - DR + F) = 0,$$

(4) $$(BR + D)^2 - 4A(CR^2 + ER + F) = 0,$$

(5) $$(-BR + D)^2 - 4A(CR^2 - ER + F) = 0,$$
respectively. Subtract (2) from (3) and divide by $4R$; this gives

(6) $$2CD - BE = 0.$$

Similarly, from (4) and (5),

(7) $$-BD + 2AE = 0.$$

By (6), (7) and (1), $D = E = 0$. Therefore (2) and (4) become

$$B^2R^2 - 4ACR^2 - 4CF = 0, \quad B^2R^2 - 4ACR^2 - 4AF = 0,$$

respectively. Since $F \neq 0$, we have $A = C$; this means that the ellipse has its axes along the lines $u \pm v = 0$.

A–5 Assume that n divides $2^n - 1$ for some $n > 1$. Since $2^n - 1$ is odd, n is odd. Let p be the smallest prime factor of n. By Euler's Theorem, $2^{\phi(p)} \equiv 1 \pmod p$, because p is odd. If λ is the smallest positive integer such that $2^\lambda \equiv 1 \pmod p$ then λ divides $\phi(p) = p - 1$. Consequently λ has a smaller prime divisor than p. But $2^n \equiv 1 \pmod p$ and so λ also divides n. This means that n has a smaller prime divisor than p Contradiction.

A–6 The conditions imply $\int_0^1 (x - \frac{1}{2})^n f(x)dx = 1$. Suppose $|f(x)| < 2^n(n+1)$ except for a set of measure 0.

Then $1 = \int_0^1 (x - \frac{1}{2})^n f(x)dx < 2^n(n+1) \int_0^1 |x - \frac{1}{2}|^n dx = 1$, a contradiction.

B–1 For the proposed solution the problem could have been stated in the more general form: The series expansion about any point for $\exp(P(x))$, if $P(x)$ is a cubic polynomial, will not have three consecutive zero coefficients.

If $f(x) = \exp(P(x))$, where $P(x)$ is a cubic polynomial, then $f' = f \cdot P'$ and $f'' = f' \cdot P' + f \cdot P''$. In general for $k \geqq 2$,

(1) $$f^{(k+1)} = f^{(k)} \cdot P' + \binom{k}{1} f^{(k-1)} \cdot P'' + \binom{k}{2} f^{(k-2)} \cdot P'''.$$

It follows from (1): if, at some (real or complex) point x_0, $f^{(k-2)}(x_0) = f^{(k-1)}(x_0) = f^{(k)}(x_0) = 0$, then also $f^{(k+1)}(x_0) = 0$. By the same argument, $f^{(\mu)}(x_0) = 0$ for $\mu = k+2, k+3, \cdots$; so that $f(x)$ would reduce to a polynomial. This is evidently impossible.

Alternate Solution: In the given form of the problem it can be shown that no coefficient of x^k is zero. The product $x^n(1-x)^{2n}$ has a non-zero coefficient for x^k if $0 \leqq k - n \leqq 2n$ or, equivalently, $k/3 \leqq n \leqq k$. This coefficient is the integer

$$(-1)^{k-n}\binom{2n}{n-k},$$

which we denote by $a(n, k)$. The coefficient of x^k in the given series is

$$C_k = \sum_{n=[k/3]+1}^{k} \frac{a(n,k)}{n!}.$$

Multiplying through this summation by $(k-1)!$ will convert each term, except the last term, to an integer. The last term becomes $1/k$. Since $(k-1)!$ times C_k is not an integer for $k > 1$ and $C_1 = C_0 = 1$, there are no zero coefficients in the expansion of the given series in powers of x.

B–2 We take v_0 as positive (see Comment) and consider the graph of v as a function of t (see Figure 1). From the given data we know that the curve starts at the origin and is concave downward since the acceleration $a = dv/dt$ does not increase.

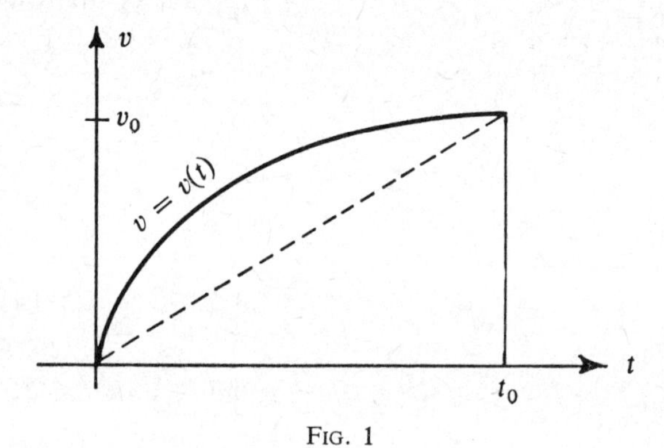

FIG. 1

Let t_0 be the time of the traverse. Then $v(t_0) = v_0$. The distance s_0 is represented by the area bounded by the curve $v = v(t)$, the t-axis, and the line $t = t_0$. The area of the right triangle with vertices at $(0,0)$, $(t_0, 0)$ and (t_0, v_0) has area less than or equal to s. Thus $\frac{1}{2} v_0 t_0 \leqq s_0$ or

$$t_0 \leqq \frac{2s_0}{v_0}.$$

Equality is possible and gives the maximum value of t_0 (for given s_0 and v_0) when the graph of $v(t)$ is the straight line $v(t) = (v_0/t_0)t = (v_0^2/2s_0)t$.

Comment: If v_0 is zero or negative, there is no maximum time t_0 for the traverse. In the case $v_0 = 0$ the equation of motion

$$S = s_0[3(t/t_0)^2 - 3(t/t_0)^3], \qquad 0 \leqq t \leqq t_0$$

satisfies the conditions of the problem for any $t_0 > 0$.

B–3 From $ABA = BA^2B = BA^{-1}B$, we have

$$AB^2 = ABA \cdot A^{-1}B = BA^{-1}BA^{-1}B = BA^{-1} \cdot ABA = B^2A.$$

By induction, $AB^{2r} = B^{2r}A$ so that $AB = AB^{2n} = B^{2n}A = BA$. Since A and B commute, $ABA = BA^2B$ implies $A^2B = A^2B^2$, or $B = B^2$, or $B = 1$.

Alternate Solution: It can be shown that A and B commute by expressing each as powers of the same group element. Because $A^3 = 1$ it is tempting to multiply $ABA = BA^2B$ on the right by A^2 and then on the left by BA^2 to get $B^2 = (BA^2)^3$. Set $X = BA^2$ and use $B^{2n} = B$ to obtain

$$B = X^{3n}. \tag{1}$$

From $X = BA^2$, we get $XA = B$, $A = X^{-1}B$, or

$$A = X^{3n-1}. \tag{2}$$

The conclusion that $B = 1$ is as before.

B–4 Let $x = t^n$, $y = t^{n+1}$, $z = t + t^{n+2}$. We construct a polynomial $P(x, y, z)$ with integral coefficients such that $P(x, y, z) = t$. We have

$$\begin{aligned} z &= t + t^{n+2}, \\ zy &= t^{n+2} + t^{2n+3}, \\ zy^2 &= t^{2n+3} + t^{3n+4}, \\ &\cdots\cdots\cdots\cdots \\ z^{n-2} &= t^{n^2-n-1} + t^{n^2}. \end{aligned}$$

Multiply the above equations alternately by $+1$ and -1 and add:

$$z[1 - y + y^2 - \cdots + (-1)^{n-2}y^{n-2}] = t + (-1)^{n-2}t^{n^2} = t + (-1)^n x^n.$$

Hence, if we define

$$P(x, y, x) = z\left[\sum_{i=0}^{n-2} (-1)^i y^i\right] + (-1)^{n-1}x^n,$$

Then $P(t^n, t^{n+1}, t + t^{n+2}) = t$.

B–5 For the skew quadrilateral $ABCD$, let $AB = a$, $BC = b$, $CD = c$, $DA = d$, $AC = x$, $BD = y$. None of these lengths can be zero. By the law of cosines:

$$\frac{a^2 + b^2 - x^2}{ab} = \frac{c^2 + d^2 - x^2}{cd}$$

or $(ab - cd)x^2 = (bc - ad)(ac - bd)$. Similarly, $(ad - bc)y^2 = (cd - ab)(ac - bd)$.

CASE 1: $ab - cd = 0$.

Then, $ad - bc = 0$ and $a = c$, $b = d$.

CASE 2: $ab - cd \neq 0$.

Then, $bc - ad \neq 0$, $ac - bd \neq 0$ and $x^2y^2 = (ac - bd)^2$. Consequently,

$$ac = xy + bd \text{ or } bd = ac + xy.$$

By Ptolemy's Theorem (in space), $ABCD$ must be concyclic which violates the skew condition.

Alternate Solution: If $AC = BC$ and $AD = BD$, the conclusion that $AC = BD$ and $BC = AD$ is obvious (see Figure 2) so assume $AC \neq BC$. With this assumption we first show $BD = AC$. If $BD \neq AC$ there exists a unique point D^* in the plane of $\triangle ADB$ with $BD^* = AC$, $AD^* = CB$. $\angle AD^*B = \angle ACB = \theta$. From $\triangle$'s ADE and BD^*E it follows that $\angle DAE = \angle D^*BE$.

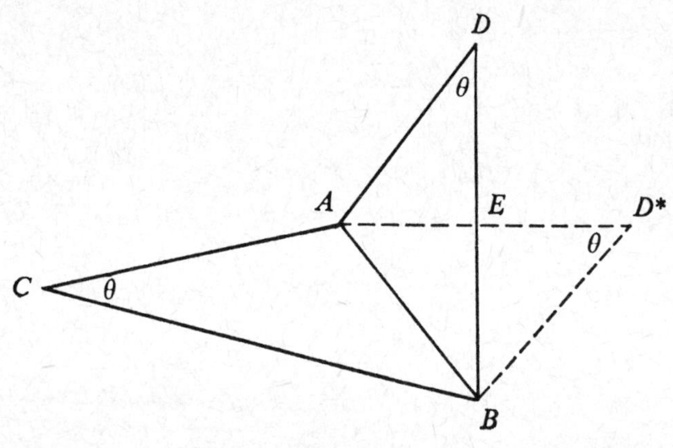

FIG. 2

From the congruent $\triangle$'s CD^*A and CD^*B it follows that $\angle CAD^* = \angle CBD^*$. These angle equalities prove that the trihedral angles $A - CDD^*$ and $B - CDD^*$ are congruent. Hence the angle which $\overrightarrow{CA}$ makes with the plane ADD^* is equal to the angle $\overrightarrow{CB}$ makes with the plane BDD^* (which is the plane ADD^*). If H is the foot of the altitude from C to this plane the $\triangle$'s CHA and CHB are congruent right triangles. That is $AC = CB$. This is a contradiction and $BD = AC$.

Interchanging the roles of B and A in the above shows that $AD = BC$.

B–6 Let $P(z)$ denote the given polynomial. The power series expansion of $1/(1 - z) - 2P(z)$ has coefficients ± 1 with leading coefficient -1. Hence,

$$(1) \qquad \left|1 + \frac{1}{1 - z} - 2P(z)\right| \leqq |z| + |z|^2 + \cdots = \frac{|z|}{1 - |z|}.$$

Also,

$$|2P(z)| \geqq \left|1 + \frac{1}{1 - z}\right| - \left|1 + \frac{1}{1 - z} - 2P(z)\right|$$

$$\geqq 1 + \frac{1}{1 + |z|} - \frac{|z|}{1 - |z|} = 2\frac{1 - |z| - |z|^2}{1 - |z|^2}.$$

The latter term is positive for $|z| < (\sqrt{5} - 1)/2$.

THE THIRTY-FOURTH WILLIAM LOWELL PUTNAM MATHEMATICAL COMPETITION

December 1, 1973

A–1. (a) If X, Y, Z are at the midpoints of the sides, the area of $\triangle XYZ$ is one fourth of the area of $\triangle ABC$. Also, as long as $\overline{BX} \leqq \overline{XC}$, $\overline{CY} \leqq \overline{YA}$ and $\overline{AZ} \leqq \overline{ZB}$, moving one of X, Y, Z to the midpoint of its side, while leaving the other two fixed, does not increase the area of $\triangle XYZ$ since the altitude to the fixed base of $\triangle XYZ$ decreases or remains constant.

(b) Under the hypothesis of (a) the three corner triangles have no more than three fourths of the total area and so one of them must have smaller area than $\triangle XYZ$. All other cases are similar to the one in which $\overline{XC} < \overline{BX}$ and $\overline{CY} < \overline{YA}$. Then consideration of the altitudes to base XY shows that $\triangle CYX$ has smaller area than $\triangle XYZ$.

A–2. The ideas in both parts are similar and the answer in (b) is "Yes." Let u_n be the nth term $\pm 1/n$ and $S_n = u_1 + \cdots + u_n$. Since $u_n \to 0$ as $n \to \infty$, $\{S_n\}$ will converge if and only if $\{S_{8m}\}$ does. Using the facts that

$$\frac{1}{n} - \frac{1}{n+k} = \frac{k}{n(n+k)},$$

that $\Sigma(1/n^2)$ converges, and that $\Sigma(1/n)$ diverges, one shows that with four "+" signs and four "−" signs in each block, $\{S_{8m}\}$ converges as the term-by-term sum of four convergent sequences while an imbalance of signs makes $\{S_{8m}\}$ divergent as the sum of a convergent and a divergent sequence.

A–3. Let $c(n) = \sqrt{4n+1}$ and let $[x]$ denote the greatest integer in x; then we wish to show that $[b(n)] = [c(n)]$. Let $k(n)$ be a value of k that minimizes $k + (n/k)$. Then

$$b(n-1) \leqq k(n) + \{(n-1)/k(n)\} < k(n) + \{n/k(n)\} = b(n),$$

i.e., $b(n-1) < b(n)$. Let m be a positive integer. Then

$$b(m^2) = 2m, b(m^2 + m) = 2m + 1. \qquad \text{(I)}$$

It follows from formulas (I) and the strictly increasing nature of $b(n)$ that

$$[b(n)] = 2m \text{ for } m^2 \leqq n < m^2 + m, \ [b(n)] = 2m + 1 \text{ for } m^2 + m \leqq n < (m+1)^2. \qquad \text{(II)}$$

On the other hand, $c(n)$ is also an increasing function and

$$c(m^2 - 1) = \sqrt{4m^2 - 3} < 2m, \ c(m^2) = \sqrt{4m^2 + 1} > 2m, \ c(m^2 + m) = \sqrt{4m^2 + 4m + 1} = 2m + 1.$$

These facts show that (II) remains true when $[c(n)]$ is substituted for $[b(n)]$.

A–4. Three; at 0, 1, and some $x > 1$. The first two are clear and the other follows from $f(4) < 0$ and $f(5) > 0$ or from $f'(1) < 0$ while $f(x) \to +\infty$ as $x \to +\infty$. There are no more zeros since four zeros of f would imply a zero of f''' using an extension of Rolle's Theorem; but $f'''(x) = (\log 2)^3 2^x \neq 0$ for all x.

A–5. (a) If two of $x(0)$, $y(0)$, $z(0)$ vanish, then $x'(0) = y'(0) = z'(0) = 0$, and the uniqueness theorem applies (the equations are clearly "Lipschitz").

(b) Clear (this was intended as a hint for part (c)).

(c) Now write the equations in the symmetric form:

$$xx' = yy' = zz' = xyz.$$

Thus $x^2 - c_1 = y^2 - c_2 = z^2 - c_3$ with constant c_i. Say without loss of generality that $c_1 \geqq c_2 \geqq c_3$, and then set $c_3 = 0$. Thus $z^2 \leqq y^2 \leqq x^2$, and:

$$z^2 = x^2 - c_1 = y^2 - c_2, \ c_i \geqq 0;$$

$$\frac{dz}{dt} = \pm \sqrt{(z^2 + c_1)(z^2 + c_2)}.$$

Now let time t move in the direction which makes $|z|$ increase (this depends on the sign of z and on the $\pm$ sign in the square root).

For simplicity assume that $z(0) \geqq 0$, and that the sign on the square root is $+$; then let time move positively. Since

$$t = \int dz / \sqrt{(z^2 + c_1)(z^2 + c_2)}$$

and the z-integral converges, a finite amount of time suffices to push z out to infinity.

A–6. Any two distinct lines in the plane meet in at most one point. There are altogether $\binom{7}{2} = 21$ pairs of lines. A triple intersection accounts for 3 of these pairs of lines, and a simple intersection accounts for 1.

Finally, $6 \cdot 3 + 4 \cdot 1 = 22 > 21$.

B–1. Since the sum of the $2n$ integers remaining is always even, no matter which of the a_i is taken away, all of the a_i must have the same parity. Now a similar argument shows that they are all congruent (mod 4); for the property held by the a_i is shared by the integers $a_i/2$ or $(a_i - 1)/2$ (depending on whether the a_i are all even or all odd). Continuing in this manner, all of the a_i are congruent (mod 2^k) for every k. This is possible for integers only if they are equal.

B–2. Let $z = e^{\theta i}$ and $z^n = w = u + iv$ (with u and v real). Then $|z^{2n} - 1| = |w^2 - 1| = [(u^2 - v^2 - 1)^2 + (2uv)^2]^{1/2} = 2|v|$, using $u^2 + v^2 = 1$. [$|z^{2n} - 1| = 2|\sin n\theta|$ is also easily shown geometrically using an isosceles triangle.] Hence it suffices to show that $v = \sin n\theta$ is rational when $x = \cos\theta$ and $y = \sin\theta$ are rational. For $n \geqq 0$, this follows from $(x + iy)^n = u + iv$ or by mathematical induction using the addition formulas for the sine and cosine. Then the case $n < 0$ follows using $\sin(-\alpha) = \sin\alpha$.

B–3. One triple (a, b, c) satisfying the conditions is $(1, -1, p)$; it remains to show that this is the only solution. Clearly b must be odd since $b^2 \equiv 1 \pmod 4$. Also $b^2 = (-b)^2$, so write $|b| = 2x - 1$. Then $b^2 - 4ac = 1 - 4p$ gives

$$x^2 - x + p = ac.$$

If $0 \leqq x < p$, the hypothesis tells us that ac is prime; then $0 < a \leqq c$ implies that $a = 1$, it follows from $-a \leqq b < a$ and the oddness of b that $b = -1$, and $1 - 4p = b^2 - 4ac = 1 - 4c$ gives us $c = p$. Since $x = (|b| + 1)/2 \geqq 0$, it suffices to show that $x < p$. Since $|b| \leqq a \leqq c$, $b^2 - 4ac = 1 - 4p$, and $p \geqq 2$, one sees that $x < p$ using

$$3a^2 = 4a^2 - a^2 \leqq 4ac - b^2 = 4p - 1,$$

$$|b| \leqq a \leqq \sqrt{(4p - 1)/3},$$

$$x = (|b| + 1)/2 < \sqrt{p/3} + (1/2) < p.$$

B–4. We give two solutions; the first does not use the hint and the second does.

THEOREM. *If f is continuous on $[0, 1]$, $f(0) = 0$, and $0 \leqq f'(x) \leqq 1$ on $(0, 1)$, then*

$$\left[\int_0^1 f(x)dx\right]^2 > \int_0^1 [f(x)]^3 dx$$

unless, identically on $[0, 1]$, either $f(x) = x$ or $f(x) = 0$.

Proof. Define $G(t) = 2\int_0^t f(x)dx - [f(t)]^2$ for $t \in [0, 1]$. Then $G(0) = 0$ and $G'(t) = 2f(t)[1 - f'(t)] \geqq 0$, so that $G(t) \geqq 0$ and consequently $f(t)G(t) \geqq 0$.

Now define $F(t) = [\int_0^t f(x)dx]^2 - \int_0^t [f(x)]^3 dx$ for $t \in [0, 1]$. Then $F(0) = 0$ and $F'(t) = f(t)G(t) \geqq 0$, so that $F(t) \geqq 0$ and in particular $F(1) \geqq 0$.

Equality is possible only if $f(t)G(t) = F'(t) = 0$ for all t, which implies that, for some K, $f = 0$ on $[0, K]$ and $G' = 0$, with $f > 0$, on $(K, 1)$. We then have $f' = 1$ on $(K, 1)$, which is admissible only if $K = 0$ or $K = 1$, since otherwise $f'(K)$ is simultaneously defined and undefined.

The unique answer to (b) is $f(x) = x$. The following is an outline of a proof of (a) using the hint. Let $f(1) = c$. The hypothesis implies that f has an inverse g with $g'(y) \geqq 1$ on $0 \leqq y \leqq c$. Let

$$A = \left[\int_0^1 f(x)dx\right]^2 \text{ and } B = \int_0^1 [f(x)]^3 dx.$$

Then

$$A = \left[\int_0^c yg'(y)dy\right]^2 = \int_0^c \int_0^c yg'(y)zg'(z)dzdy = 2\int_0^c \int_0^z yg'(y)zg'(z)dydz$$

using the symmetry of the integrand about the line $y = z$. Now $g'(y) \geqq 1$ implies

$$A \geqq \int_0^c zg'(z) \left[\int_0^z 2ydy\right]dz = \int_0^c z^3 g'(z)dz = B.$$

B–5. (a) Let $r = -b/a$ and $s = -c/a$. Let polynomials p_n and q_n in r and s be defined by the initial conditions $p_0 = 0$, $p_1 = 1$, $q_0 = 1$, and $q_1 = 0$ and the recursion formulas $p_n = rp_{n-1} + sp_{n-2}$ and $q_n = rq_{n-1} + sq_{n-2}$ for $n > 1$. Using $z^n = rz^{n-1} + sz^{n-2}$ and mathematical induction, one proves that $z^n = p_n z + q_n$ and that all the coefficients in $p_n(r, s)$ are positive. Then multiplying numerator and denominator of the right hand side of $z = [z^n - q_n(-b/a, -c/a)]/p_n(-b/a, -c/a)$ by the proper power of a leads to $z = F(z^n, a, b, c)/G(a, b, c)$, where F and G are polynomials with integer coefficients. Since all the coefficients in $p_n(r, s)$ are positive, the same is true of $G(a, b, c)$. Therefore $G(a, b, c)$ is not identically zero and F/G is the desired rational function.

(b) Let $v = x + (1/x)$. Then $x^2 - vx + 1 = 0$. Using (a) with z replaced by x, one finds that $x^3 = p_3 x + q_3$ with $p_3 = v^2 - 1$ and $q_3 = -v$. Then

$$x = (x^3 - q_3)/p_3 = (x^3 + v)/(v^2 - 1).$$

B–6. (a) Simple calculus.

(b) By induction: The case $n = 1$ is just (a).

Now the ratio of the expression for $n+1$ to the expression for n is equal to:

$$\left|\sin^2 2^n\theta \cdot \sin 2^{n+1}\theta\right|.$$

Since $\theta = \pi/3$ gives $2^n\theta \equiv 2\pi/3$ or $4\pi/3 \pmod{2\pi}$, this ratio is maximized at $\theta = \pi/3$, and by induction, then, the whole expression is maximized.

(c) Set $\theta = \pi/3$, and observe that the expression in part (b) is then exactly equal to $(3/4)^{3n/2}$; its 2/3 power is thus equal to $(3/4)^n$. That is the maximum; in general the 2/3 power of the expression in (b) is $\leqq (3/4)^n$. To get from that to the expression in (c), we would increase the powers of the end factors $\sin\theta$ and $\sin 2^n\theta$; this can only decrease the product, since $|\sin\theta| \leqq 1$.

THE THIRTY-FIFTH WILLIAM LOWELL PUTNAM MATHEMATICAL COMPETITION

December 7, 1974

A–1.

A conspiratorial subset (CS) of $\{1,2,\cdots,16\}$ has at most two numbers from the pairwise relatively prime set $\{1,2,3,5,7,11,13\}$ and so has at most $16-(7-2)=11$ numbers. But

$$\{2, 3, 4, 6, 8, 9, 10, 12, 14, 15, 16\}$$

is a CS with 11 elements; hence the answer is 11.

A–2.

Let C be the other point of intersection of line AB with the circle and let θ be the inclination of AB. Let $\overline{AB}=b$ and $\overline{AC}=c$. The square of the time of descent is proportional to $b/\sin\theta$ and hence to $1/(c\sin\theta)$, since it is well known that bc is constant with respect to θ. The time is minimized by maximizing $c\sin\theta$; this is done by choosing C as the bottom of the circle.

A–3.

If $p\equiv 1 \pmod 4$, either (A): $p\equiv 1 \pmod 8$ or (B): $p\equiv 5 \pmod 8$. We show that (A) and (B) are necessary and sufficient for (a) and (b), respectively. If $p=m^2+n^2$ and p is odd, one can let m be odd and n be even. Then $p=m^2+4v^2$ with $m^2\equiv 1 \pmod 8$. With (A), v is even and $p=m^2+16w^2$. Conversely, $p=m^2+16w^2$ implies $p\equiv m^2\equiv 1 \pmod 8$. With (B), v is odd, $m=2u+v$ for some integer u, and $p=(2u+v)^2+4v^2=4u^2+4uv+5v^2$. Conversely, $p=4u^2+4uv+5v^2$ with p odd implies $p=(2u+v)^2+4v^2$ with v odd and hence $p\equiv 5 \pmod 8$.

A–4.

The answer is

$$\frac{n}{2^{n-1}}\binom{n-1}{[(n-1)/2]}$$

since

$$\begin{aligned}\sum_{k<n/2}(n-2k)\binom{n}{k} &= \sum_{k<n/2}\left\{(n-k)\binom{n}{k}-k\binom{n}{k}\right\}\\ &= \sum_{k<n/2}\left\{n\binom{n-1}{k}-n\binom{n-1}{k-1}\right\} = n\sum_{k<n/2}\left\{\binom{n-1}{k}-\binom{n-1}{k-1}\right\}\\ &= n\binom{n-1}{[(n-1)/2]}.\end{aligned}$$

A–5.

Let F be the fixed focus, M be the moving focus, and T be the (varying) point of mutual tangency. The reflecting property of parabolas tells us that the tangent line at T makes equal angles with FT and with a vertical line. This and congruence of the two parabolas imply that MT is vertical and that the segments $\overline{FT}$ and $\overline{MT}$ are equal. Now M must be on the horizontal fixed directrix $y = 1/4$ by the focus-directrix definition of a parabola.

A–6.

Let $p(k, x)$ be the monic polynomial $(x+1)(x+2)\cdots(x+k)$ and let m be an integer. Then $p(k, m)$ is exactly divisible by $k!$ since the absolute value of the quotient is a binomial coefficient (even when m is negative). Hence, if $n \mid k!$ there is a monic integral polynomial $f(x)$ of degree k with $n \mid f(m)$ for all integers m. Conversely, the condition $n \mid k!$ is necessary since the k-th difference $k!$ of a monic integral polynomial of degree k is divisible by any common divisor of all the values $f(m)$.

In particular, $k(10^6) = k(5^6 2^6) = 25$ since the smallest s with $5^6 \mid s!$ is $s = 25$.

B–1.

Since the p_i need not be distinct, the sum is a continuous function on the compact set $C \times C \times C \times C \times C$, where C is the circle. Hence maxima exist. One proves that the maximum occurs when the p_i are the vertices of a regular pentagon by showing that this configuration simultaneously maximizes both of the sums:

$$S = d(p_1, p_2) + d(p_2, p_3) + d(p_3, p_4) + d(p_4, p_5) + d(p_5, p_1),$$

$$T = d(p_1, p_3) + d(p_2, p_4) + d(p_3, p_5) + d(p_4, p_1) + d(p_5, p_2).$$

For S or T, one can fix four of the points; then the varying part of the sum is of the form.

$$D = d(p, a) + d(p, b), \text{ with } a \text{ and } b \text{ fixed}.$$

Using the Law of Sines, one shows that D is a constant times $\sin\alpha + \sin\beta$ where $\alpha = \measuredangle pab$, $\beta = \measuredangle pba$, and $\alpha + \beta$ is constant. Then it is easy to show that D is not a maximum unless p is symmetrically situated with respect to a and b.

B–2.

If $y'(x_n) = 0$ for a sequence $\{x_n\}$ approaching $+\infty$, the hypothesis insures that $y(x_n) \to 0$. Since these x_n may include any relative maxima and minima, this case must have $y(x) \to 0$ as $x \to +\infty$. Then one also has $y'(x) \to 0$ as $x \to +\infty$.

In the remaining case, there is an x_0 such that for $x > x_0$ one has $y' \neq 0$ and so $(y')^2 > 0$. We restrict ourselves to the x's with $x > x_0$ and consider two subcases:

(a) $y' > 0$. If y is unbounded above, so are y^3 and $(y')^2 + y^3$. This contradicts the hypothesis $(y')^2 + y^3 \to 0$ as $x \to +\infty$. If y is bounded above, it approaches a finite limit. Then y^3, $(y')^2$, and y' approach limits. Since y is bounded, the limit for y' must be 0. Then y also has 0 as its limit.

(b) $y' < 0$. There is no problem unless y is unbounded below. Then we may assume that $y < 0$ and compare y to a solution of the differential equation

$$y' = -(1/2)|y|^{3/2}, \qquad y < 0.$$

Every solution diverges to $-\infty$ in a finite interval, hence so does $y(x)$; this contradicts the hypothesis that y is defined and smooth for all large x.

B–3.

If $\alpha = r/s$ with r and s integers and $s > 0$, then $\cos(n\pi\alpha)$ takes on at most $2s$ distinct values for integral choices of n. When $\cos \pi\alpha = 1/3$, the formula $\cos 2\theta = 2\cos^2\theta - 1$ and mathematical induction can be used to show that

$$\cos(2^m \pi\alpha) = t/3^{2^m - 1} \qquad [m = 1, 2, 3, \cdots],$$

with t an integer not divisible by 3, and hence that these cosines form an infinite set of distinct values. Thus α is irrational.

B–4.

For each n, we construct the function $g_n(x, y)$ as follows: First divide the xy-plane into vertical strips of width $1/n$ separated by the lines $\{x = m/n\}$, m an integer. Now set $g_n(x, y) = f(x, y)$ along each vertical line $x = m/n$, and interpolate linearly (holding y fixed and letting x vary) in between. Then $g_n(x, y)$ is continuous because $f(x_0, y)$ is continuous in y; $g_n(x, y) \to f(x, y)$ because $f(x, y_0)$ is continuous in x.

REMARKS. This result has two interesting consequences for functions which are continuous in each variable separately:

(i) Such functions are Borel measurable.

(ii) They are continuous (in the usual sense) except on a set of points of the first Baire category. (In particular, there is no function which is continuous in each variable separately and yet discontinuous at every point.)

B–5.

We want to show that

$$\sum_{k=0}^{n} \frac{n^k}{k!} = e^n - \frac{1}{n!}\int_0^n (n-t)^n e^t \, dt > \frac{e^n}{2}$$

or, equivalently, that

$$n! > 2e^{-n}\int_0^n (n-t)^n e^t \, dt,$$

$$\int_0^\infty t^n e^{-t} \, dt > 2e^{-n}\int_0^n (n-t)^n e^t \, dt.$$

Letting $u = n - t$, this can be transformed into

$$\int_0^\infty t^n e^{-t} \, dt > 2\int_0^n u^n e^{-u} \, du,$$

which is equivalent to

$$\int_n^\infty u^n e^{-u} \, du > \int_0^n u^n e^{-u} \, du.$$

Let $f(u) = u^n e^{-u}$. Then it suffices to show that

$$f(n+h) \geqq f(n-h) \quad \text{for} \quad 0 \leqq h \leqq n.$$

This is equivalent to

$$(n+h)^n e^{-h} \geqq (n-h)^n e^h,$$

$$n \ln(n+h) - h \geqq n \ln(n-h) + h.$$

Let $g(h) = n \ln(n+h) - n \ln(n-h) - 2h$. Then $g(0) = 0$ and

$$\frac{dg}{dh} = \frac{n}{n+h} + \frac{n}{n-h} - 2 = \frac{2n^2}{n^2 - h^2} - 2 > 0$$

for $0 < h < n$. Hence $g(h) > 0$ for $0 < h < n$. The desired result follows.

B–6.

Let $n \equiv r \pmod 6$ with r in $\{0, 1, 2, 3, 4, 5\}$. Then the pattern is

r	0	1	2	3	4	5
$s_{0,n}$	$a+1$	b	c	$d-1$	e	f
$s_{1,n}$	a	b	$c+1$	d	e	$f-1$
$s_{2,n}$	a	$b-1$	c	d	$e+1$	f

This is easily proved by mathematical induction using the formulas

$$s_{i,n} = s_{i-1,n-1} + s_{i,n-1}. \qquad [\text{Here } 0-1 = 2 \pmod 3).]$$

These formulas follow immediately from the rule

$$\binom{n}{k} = \binom{n-1}{k-1} + \binom{n-1}{k}.$$

The sums may be computed readily using the above patterns and

$$s_{0,n} + s_{1,n} + s_{2,n} = 2^n.$$

For $n = 1000$, $r = 4$ and

$$s_{0,1000} = s_{1,1000} = s_{2,1000} - 1 = (2^{1000} - 1)/3.$$

THE THIRTY-SIXTH WILLIAM LOWELL PUTNAM MATHEMATICAL COMPETITION

December 6, 1975

A–1.

Let $n = [(a^2 + a)/2] + [(b^2 + b)/2]$, with a and b integers. Then

$$4n + 1 = 2a^2 + 2a + 2b^2 + 2b + 1 = (a + b + 1)^2 + (a - b)^2.$$

Conversely, let $4n + 1 = x^2 + y^2$, with x and y integers. Then exactly one of x and y is odd and so $a = (x + y - 1)/2$ and $b = (x - y - 1)/2$ are integers. One easily verifies that

$$[(a^2 + a)/2] + [(b^2 + b)/2] = (x^2 + y^2 - 1)/4 = n.$$

A–2.

The desired region is the inside of the triangle with vertices $(0, -1)$, $(2, 1)$, $(-2, 1)$. The boundary segments lie on the lines

$$\mathrm{L}_1\colon \mathrm{c} = 1, \mathrm{L}_2\colon \mathrm{c} - \mathrm{b} + 1 = 0, \mathrm{L}_3\colon \mathrm{c} + \mathrm{b} + 1 = 0.$$

To see this, we let $f(z) = z^2 + bz + c$ and denote its zeros by r and s. Then $-b = r + s$ and $c = rs$. Also

$$(r + 1)(s + 1) = rs + r + s + 1 = c - b + 1 = f(-1),$$
$$(r - 1)(s - 1) = rs - r - s + 1 = c + b + 1 = f(1).$$

On or below L_2, at least one zero is real and not greater than -1; this follows either from $(r + 1)\cdot(s + 1) \leqq 0$ or from $f(-1) \leqq 0$ and the fact that the graph of $y = f(x)$, for x real, is an upward opening parabola. Similarly, on or below L_3 one zero is real and at least 1. On or above L_1, at least one zero has absolute value greater than or equal to 1. Hence the desired points (b, c) must be inside the described triangle.

Conversely, if (b, c) is inside the triangle, $|c| < 1$ and so $|r| < 1$ or $|s| < 1$ or both. If the zeros are complex, they are conjugates and $|r| = |s|$; then $|r| = |s| < 1$ follows from $|c| < 1$. If the zeros are real, $|c| < 1$ implies that at least one zero is in $(-1, 1)$. Then $(r + 1)(s + 1) = f(-1) > 0$ and $(r - 1)\cdot(s - 1) = f(1) > 0$ imply that the other zero is also in $(-1, 1)$.

For full credit, the region had to be depicted.

A–3.

Let $h(x) = x^a - x^b$ and $k(z) = z^c - z^b$. The desired points also give the maximum and minimum of the function

$$g(x, z) = (x^a + y^b + z^c) - (x^b + y^b + z^b) = h(x) + k(z)$$

on the domain obtained by projection of the solid domain on the xz-plane. For all points under consideration, both x and z are in $[0,1]$. Examining its derivative, one sees that $h(x)$ increases from 0 at $x=0$ to a maximum at $x_0=(a/b)^{1/(b-a)}$ and then decreases to 0 at $x=1$. (This uses the hypothesis $0<a<b$.) Similarly, $k(z)$ decreases from 0 at $z=0$ to a minimum at $z_0=(b/c)^{1/(c-b)}$ and then increases to 0 at $z=1$. Since $(1, z_0)$ and $(x_0, 1)$ are not in the domain of $g(x,z)$, the function f achieves its maximum only at $(x,y,z)=(x_0,[1-x_0^b]^{1/b},0)$ and achieves its minimum only at $(0,[1-z_0^b]^{1/b},z_0)$.

A-4.

Let $n=4k+2$ with $k>0$. Then

$$0=\theta^n-1=\theta^{4k+2}-1=(\theta^{2k+1}-1)(\theta^{2k+1}+1),$$
$$0=(\theta^{2k+1}-1)(\theta+1)(\theta^{2k}-\theta^{2k-1}+\theta^{2k-2}-\cdots-\theta+1).$$

Since θ is a primitive nth root of unity with $n>2k+1$ and $n>2$,

$$(\theta^{2k+1}-1)(\theta+1)\neq 0.$$

Hence

(A) $$\theta^{2k}-\theta^{2k-1}+\theta^{2k-2}-\cdots+\theta^2-\theta+1=0,$$
$$1=\theta-\theta^2+\theta^3-\cdots-\theta^{2k}=(1-\theta)(\theta+\theta^3+\theta^5+\cdots+\theta^{2k-1}),$$
$$(1-\theta)^{-1}=\theta+\theta^3+\cdots+\theta^{2k-1}\ [\text{where } 2k-1=(n-4)/2].$$

Another solution is $(1-\theta)^{-1}=1+\theta^2+\theta^4+\cdots+\theta^{2k}$ as one sees from (A).

A-5.

The answer for c is $\sqrt{2/w}$, where w is the wronskian $y_1y_2'-y_2y_1'$ (and will be seen below to be constant).

Let $c^2=2k$. Then $z^2/2=ky_1y_2$. Differentiating twice, one has

$$zz'=k(y_1y_2'+y_2y_1'),\ zz''+(z')^2=k(y_1y_2''+y_2y_1''+2y_1'y_2').$$

Since $y_1''=fy_1$ and $y_2''=fy_2$, this implies

$$zz''+(z')^2=2k(fy_1y_2+y_1'y_2')=f(2ky_1y_2)+2ky_1'y_2'=fz^2+2ky_1'y_2'.$$

Now

$$z^3z''+(zz')^2=fz^4+2kz^2y_1'y_2',$$
$$z^3z''+k^2(y_1y_2'+y_2y_1')^2=fz^4+4k^2(y_1y_2y_1'y_2')$$
$$z^3z''+k^2(y_1y_2'-y_2y_1')^2=fz^4,$$

(1) $$z^3z''-fz^4=-k^2(y_1y_2'-y_2y_1')^2=-k^2w^2=-c^4w^2/4$$

Since $w'=(y_1y_2'-y_2y_1')'=y_1y_2''-y_2y_1''=y_1(fy_2)-y_2(fy_1)=0$, w is a constant. Solving $c^4w^2/4=1$ for c gives $c=\sqrt{2/w}$; for this c, (1) implies $z''-fz=-z^{-3}$ or $z''+z^{-3}=fz$.

A-6.

Let λ denote the line through the desired points P_4 and P_5. Let π be the plane of P_1, P_2, and P_3 and let H be the intersection of λ with π.

Let v_u be the vector $\mathbf{HP}_u$ and $|v_u|$ be its magnitude. We wish to have the dot product

(1) $$d=\mathbf{P}_h\mathbf{P}_k\cdot\mathbf{P}_i\mathbf{P}_j=(v_k-v_h)\cdot(v_j-v_i)=v_k\cdot v_j-v_k\cdot v_i-v_h\cdot v_j+v_h\cdot v_i$$

zero for all choices of h, k, i, j as distinct indices in $\{1,2,3,4,5\}$.

Since λ is to be perpendicular to π, we must have

(2) $$v_h \cdot v_i = 0 \text{ for } h \in \{4,5\} \text{ and } i \in \{1,2,3\}.$$

If $h, k \in \{4,5\}$ and $i, j \in \{1,2,3\}$, (2) implies that the dot product d of (1) is zero. If $h \in \{4,5\}$ and $k, i, j \in \{1,2,3\}$, (2) implies that the d of (1) becomes

(3) $$d = v_k \cdot v_j - v_k \cdot v_i = v_k \cdot (v_j - v_i) = \mathbf{HP}_k \cdot \mathbf{P}_i\mathbf{P}_j.$$

Clearly the d of (3) are zero simultaneously if and only if H is the orthocenter (i.e., intersection of altitudes) of $\Delta P_1P_2P_3$. With this choice of H, the vanishing of the d of (3) implies

(4) $$v_2 \cdot v_3 = v_1 \cdot v_3 = v_1 \cdot v_2.$$

Now let $h, i \in \{4,5\}$ and $k, j \in \{1,2,3\}$. Then (2) implies

(5) $$d = v_k \cdot v_j + v_4 \cdot v_5.$$

Assuming (4), one sees that all the d's of (5) will be zero if $v_4 \cdot v_5 = -v_1 \cdot v_2$. The hypothesis that $\Delta P_1P_2P_3$ is acute-angled tells us that H is inside the triangle. Then at least one (actually, all) of the angles $\measuredangle P_1HP_2$, $\measuredangle P_2HP_3$, $\measuredangle P_3HP_1$ must be obtuse and so the equal dot products of (4) must be negative. Hence $v_4 \cdot v_5$ must be positive; this means that P_4 and P_5 must be on the same half-line of λ determined by H.

Now the location of P_4 and P_5 can be given. Let H be the orthocenter of $\Delta P_1P_2P_3$ and μ be either half-line perpendicular to plane $P_1P_2P_3$ at H. Then P_4 may be any point on μ such that $|v_4|$ is neither zero nor $(-v_1 \cdot v_2)^{1/2}$ and P_5 must be the unique point on μ with $|v_5| = -v_1 \cdot v_2/|v_4|$. Then each d of (1) is zero and no three of the P_i are collinear.

B–1.

The answer is $a = 7$. Also one must have $b \equiv 5 \pmod 7$.

Proof: The subgroup H must contain $4(3,8) - 3(4,-1) = (0,35)$, $4(5,4) - 5(4,-1) = (0,21)$, and then $2(0,21) - (0,35) = (0,7)$. Now $(0,7)$ and $(1,b)$ will generate H iff $(1,b)$ is in H and there exist integers u, v, and w such that

$$(3,8) = 3(1,b) + u(0,7),\ (4,-1) = 4(1,b) + v(0,7),\ (5,4) = 5(1,b) + w(0,7).$$

These hold iff $8 = 3b + 7u$, $-1 = 4b + 7v$, and $4 = 5b + 7w$. With $b = 5 + 7k$, k any integer, the desired coefficients u, v, and w exist in the form $u = -1 - 3k$, $v = -3 - 4k$, $w = -3 - 5k$. It now suffices to let $k = 0$ and to note that $(1,5) = (4,-1) - (3,8) + 2(0,7)$ is in H.

B–2.

Let $\Sigma d_i = d$ and let S be a sphere of radius $r > d/2$. The area of S contained in slab S_i is at most $2\pi d_i$. It follows that the area of S contained in the union of the slabs S_i is at most $2\pi d < 4\pi r$ = (area of S). Hence there are points of S that are not in any of the slabs.

The problem may also be done using volumes of intersection of the slabs with an appropriately chosen sphere.

B–3.

In the expansion of $s_1^k = (a_1 + a_2 + \cdots + a_n)^k$, every term of s_k appears with $k!$ as coefficient and the other coefficients are nonnegative. Hence $s_k/s_1^k \leqq 1/k!$

If we let each $a_i = 1$,

$$\frac{s_k}{s_1^k} = \binom{n}{k} \bigg/ n^k = \frac{n(n-1)\cdots(n-k+1)}{k!n^k} = \frac{1}{k!}\left(1 - \frac{1}{n}\right)\left(1 - \frac{2}{n}\right)\cdots\left(1 - \frac{k-1}{n}\right),$$

which approaches $1/k!$ as k is held fixed and n goes to infinity. These facts show that the supremum M_k is $1/k!$.

B–4.

No. Since the mapping with $(x, y) \to (-x, -y)$ is a homeomorphism of the unit circle on itself, the complement $-B$ of such a subset B would also be closed. Thus the existence of such a B would make C the union $-B \cup B$ of disjoint nonempty closed subsets; this would contradict the fact that C is connected.

B–5.

Since $f_0(x) = \sum_{k=0}^{\infty} x^k/k!$, one easily shows by mathematical induction that $f_n(x) = \sum_{k=0}^{\infty}(k^n x^k/k!)$. Then, since all terms are positive, one has

$$\sum_{n=0}^{\infty} \frac{f_n(1)}{n!} = \sum_{n=0}^{\infty}\sum_{k=0}^{\infty} \frac{k^n}{k!\,n!} = \sum_{k=0}^{\infty} \frac{1}{k!}\sum_{n=0}^{\infty}\frac{k^n}{n!} = \sum_{k=0}^{\infty}\frac{e^k}{k!} = e^e.$$

B–6.

Both parts are done easily using the Inequality on the Means. For (a), one has

$$\frac{n+s_n}{n} = \frac{(1+1)+(1+\frac{1}{2})+\cdots+(1+(1/n))}{n} > \sqrt[n]{(1+1)(1+\tfrac{1}{2})\cdots(1+(1/n))} = \sqrt[n]{2\cdot\tfrac{3}{2}\cdot\tfrac{4}{3}\cdots(n+1)/n}$$
$$= (n+1)^{1/n}$$

and so $n + s_n > n(n+1)^{1/n}$.

For (b), one has

$$\frac{n-s_n}{n-1} = \frac{(1-\frac{1}{2})+(1-\frac{1}{3})+\cdots+(1-(1/n))}{n-1} > \sqrt[n-1]{(1-\tfrac{1}{2})(1-\tfrac{1}{3})\cdots(1-(1/n))} = \sqrt[n-1]{\tfrac{1}{2}\cdot\tfrac{2}{3}\cdots(n-1)/n}$$
$$= n^{-1/(n-1)}$$

and so $n - s_n > (n-1)n^{-1/(n-1)}$.

THE THIRTY-SEVENTH WILLIAM LOWELL PUTNAM MATHEMATICAL COMPETITION

December 4, 1976

A–1.

Let μ be the angle bisector of $\measuredangle AOB$ and λ be the perpendicular to μ through P. Then the intersections of λ with **OA** and **OB** are chosen as X and Y respectively.

This construction makes $OX = OY$ and there is a circle Γ tangent to **OA** at X and to **OB** at Y. Let $\overline{X_1Y_1}$ be any other segment containing P with X_1 on **OA** and Y_1 on **OB**. Let X_2 and Y_2 be the intersections of $\overline{X_1Y_1}$ with Γ. A theorem of Euclidean geometry states that $(PX)(PY) = (PX_2)(PY_2)$. Clearly $(PX_2)(PY_2)$ is less than $(PX_1)(PY_1)$. Hence $(PX)(PY)$ is a minimum.

One can also locate X and Y by saying that $(\pi - \measuredangle AOB)/2$ should be chosen as the measure of $\measuredangle OXP$ or $\measuredangle OYP$.

A–2.

One easily verifies that

$$(x+y)^n = (x+y)^{n-2}Q + (x+y)^{n-3}P,$$

$$x^n + y^n = (x^{n-2} + y^{n-2})Q - (x^{n-3} + y^{n-3})P.$$

Subtracting or adding corresponding sides gives

(R) $$F_n = QF_{n-2} + PG_{n-3},\ G_n = QG_{n-2} + PF_{n-3}.$$

The desired results now follow by strong mathematical induction using the given results for G_2, F_3, G_4, F_5, and G_6 and (R).

A–3.

We show that the only solutions are given by $3^2 - 2^3 = 1$, i.e., $(p, r, q, s) = (3, 2, 2, 3)$ or $(2, 3, 3, 2)$.

Clearly either p or q is 2. Suppose $q = 2$. Then p is an odd prime with $p^r \pm 1 = 2^s$. If r is odd, $(p^r \pm 1)/(p \pm 1)$ is the odd integer $p^{r-1} \mp p^{r-2} + p^{r-3} \mp p^{r-4} + \cdots + 1$, which is greater than 1 since $r > 1$; this contradicts the fact that 2^s has no such factor.

Now we try r as an even integer $2t$. Then $p^r + 1 = 2^s$ leads to

$$2^s = (p^t)^2 + 1 = (2n+1)^2 + 1 = 4n^2 + 4n + 2,$$

which is impossible since $4 \mid 2^s$ for $s > 1$ and $4 \nmid (4n^2 + 4n + 2)$.

Also $r = 2t$ and $p^r - 1 = 2^s$ leads to $(p^t)^2 - 1 = (2n+1)^2 - 1 = 4n^2 + 4n = 4n(n+1) = 2^s$. Since either n or $n + 1$ is odd, this is only possible for $n = 1$, $s = 3$, $p = 3$, and $r = 2$.

A–4.

We show that one answer is $s = -1/(r+1)$ and another answer is $s = -(r+1)/r = -1 - (1/r)$. Since $P(x)$ is irreducible, so is $M(x) = P(x-1)$. Hence $M(x)$ is the only monic cubic over the rationals with $r + 1$ as a zero, i.e., $M(x) = x^3 + cx^2 + dx + 1$. If the zeros of P are r, s, and t, the zeros of

M are $r+1$, $s+1$, and $t+1$. Now the coefficients -1 and 1 of x^0 in P and M, respectively, tell us that $rst=1$ and $(r+1)(s+1)(t+1)=-1$. Then

$$st=\frac{1}{r}, s+t=(s+1)(t+1)-st-1=-\frac{1}{r+1}-\frac{1}{r}-1=-\frac{r^2+3r+1}{r(r+1)}.$$

Hence s is either root of

$$x^2+\frac{r^2+3r+1}{r(r+1)}x+\frac{1}{r}=0.$$

Using the quadratic formula, one finds that s is $-1/(r+1)$ or $-(r+1)/r$.

A–5.

It is shown below that $a=2\pi$, $b=1$, and $c=1$. We use $I[S]$ to denote the integral of $e^{-D(x,y)}$ over a region S. Since $D(x,y)=0$ on R, $I[R]=A$. Now let σ be a side of R, s be the length of σ, and $S(\sigma)$ be the half strip consisting of the points of the plane having a point on σ as the nearest point of R. Changing to (u,v)-coordinates with u measured parallel to σ and v measured perpendicular to σ, one finds that $I[S(\sigma)]=\int_0^s\int_0^\infty e^{-v}dvdu=s$. The sum Σ_1 of these integrals for all the sides of R is L.

If v is a vertex of R, the points with v as the nearest point of R lie in the inside $T(v)$ of an angle bounded by the rays from v perpendicular to the edges meeting at v; let $\alpha=\alpha(v)$ be the measure of this angle. Using polar coordinates, one has

$$I[T(v)]=\int_0^\alpha\int_0^\infty re^{-r}drd\theta=\alpha.$$

The sum Σ_2 of the $I[T(v)]$ for all vertices v of R is 2π. Now the original double integral equals $\Sigma_2+\Sigma_1+A=2\pi+L+A$. Hence $a=2\pi$ and $b=1=c$.

A–6.

Let $G(x)=[f(x)]^2+[f'(x)]^2$ and $H(x)=f(x)+f''(x)$. Since H is continuous, it suffices to show that H changes sign. We assume that either $H(x)>0$ for all x or $H(x)<0$ for all x and obtain a contradiction.

Since $|f(0)|\leqq 1$ and $G(0)=4$, either $f'(0)\geqq\sqrt{3}$ or $f'(0)\leqq-\sqrt{3}$. We deal with the case in which $H(x)>0$ for all x and $f'(0)\geqq\sqrt{3}$; the other cases are similar.

Assume that the set S of positive x with $f'(x)<1$ is nonempty and let g be the greatest lower bound of S. Then $f'(0)\geqq\sqrt{3}$ and continuity of $f'(x)$ imply $g>0$. Now $f'(x)\geqq 0$ and $H(x)\geqq 0$ for $0\leqq x\leqq g$ lead to

$$G(g)=4+2\int_0^g f'(x)[f(x)+f''(x)]\,dx\geqslant 4.$$

Since $|f(g)|\leqq 1$, this implies $f'(g)\geqq\sqrt{3}$. Then continuity of $f'(x)$ tells us that there is an $a>0$ such that $f'(x)\geqq 1$ for $0\leqq x<g+a$. This contradicts the definition of g and hence S is empty. Now $f'(x)\geqq 1$ for all x and this implies that $f(x)$ is unbounded, contradicting $|f(x)|\leqq 1$. This contradiction means that $H(x)$ must change sign and so $H(x_0)=0$ for some real x_0.

Alternately, we use the Mean Value Theorem to deduce the existence of a and b with $-2<a<0<b<2$ and

$$|f'(a)|=\frac{|f(0)-f(-2)|}{2}\leqq\frac{|f(0)|+|f(-2)|}{2}\leqq\frac{1+1}{2}=1$$

and similarly $|f'(b)|\leqq 1$. Then $G(a)=[f(a)]^2+[f'(a)]^2\leqq 1+1=2$ and also $G(b)\leqq 2$. Since $G(0)=4$, $G(x)$ attains its maximum on $a\leqq x\leqq b$ at an interior point x_0 and hence $G'(x_0)=f'(x_0)H(x_0)=0$. But $f'(x_0)\neq 0$ since otherwise $[f(x_0)]^2=G(x_0)\geqq 4$ and $|f(x_0)|>1$. Thus $H(x_0)=0$.

B–1.

It is shown below that $a = 4$ and $b = 1$. Let $f(x) = [2/x] - 2[1/x]$. Then the desired limit L equals $\int_0^1 f(x)dx$. For $n = 1, 2, \ldots, f(x) = 0$ on $2/(2n+1) < x \leqq 1/n$ and $f(x) = 1$ on $1/(n+1) < x \leqq 2/(2n+1)$. Hence

$$L = \left(\frac{2}{3} - \frac{2}{4}\right) + \left(\frac{2}{5} - \frac{2}{6}\right) + \cdots = -1 + 2\left(1 - \frac{1}{2} + \frac{1}{3} - \cdots\right)$$

$$= -1 + 2\int_0^1 \frac{dx}{1+x} = -1 + 2\ln 2 = \ln 4 - 1.$$

B–2.

The answers are (a) 8; (b) $1, A^2, B, B^2, B^3, B^4, B^5, B^6$. Since $B = (B^4)^2$, $B^3 = (B^5)^2$, $B^5 = (B^6)^2$, the elements in the answer to (b) are all squares in G. They are distinct since B has order 7 and A has order 4. To show that there are no other squares, we first note that $ABA^{-1}B = 1$ implies $AB = B^{-1}A$. Then

$$AB^2 = (B^{-1}A)B = B^{-1}(AB) = B^{-1}(B^{-1}A) = B^{-2}A.$$

Similarly $AB^n = B^{-n}A$ for the other n's in $\{0, 1, \ldots, 6\}$ and so for all integers n. With this, one obtains

$$\text{(P)} \qquad (B^iA^j)(B^hA^k) = B^uA^v \text{ with } u = i + (-1)^jh,\ v = j + k.$$

Thus the set S of elements of the form B^iA^j is closed under multiplication. S is finite since i and j may be restricted to $0 \leqq i \leqq 6$ and $0 \leqq j \leqq 3$. Hence S is a group and so $S = G$. It then follows from (P) that the squares in G are the B^uA^v with $u = i[1 + (-1)^j]$ and $v = 2j$. If j is odd, $u = 0$ and $v \equiv 2 \pmod 4$. If j is even, $v \equiv 0 \pmod 4$. Thus there are no squares other than those listed above.

B–3.

The statement to be proved is false for $n \geqq 5$ unless the hypothesis is strengthened to state that A_i is independent of the conjunction of $A_1, A_2, \ldots, A_{i-2}$ for $3 \leqq i \leqq n$.

The following counterexample with $n = 5$ was furnished by Professor David M. Bloom of Brooklyn College. Let $h = 33/37$ and $k = 1/(64 + h)$. Let $P(A_i)$ be the sum of the numbers in the second row of the following table for which A_i appears in the heading:

$A_1A_2A_4A_5$	A_4	$A_1A_3A_4A_5$	$A_1A_2A_3A_4A_5$	$A_1A_2A_3A_4$	$A_2A_3A_4A_5$
$12k$	$3k$	$6k$	$7k$	$12k$	$6k$

$A_1A_3A_5$	$A_1A_2A_3A_5$	$A_2A_3A_5$	$A_3A_4A_5$
$3k$	$9k$	$3k$	$3k$

Then each $P(A_i)$ is $49k$ and $P(A_i \wedge A_j) = 37k$ for all i, j with $|i - j| > 1$. Since $(49k)^2 = 37k$, the original independence hypothesis holds. Also, $P(A_i) = 1 - a$, where $a = (15 + h)k < 1/4$. However, for any $a \leqq 1/4$, we have $u_5 \geqq 7/64$ and $P(A_1 \wedge A_2 \wedge A_3 \wedge A_4 \wedge A_5) = 7k < 7/64$.

B–4.

We let $P = (x, y)$ and the ellipse have the equation $b^2x^2 + a^2y^2 = a^2b^2$, with $a > b > 0$. Then $F_1 = (-c, 0)$ and $F_2 = (c, 0)$ with $c^2 = a^2 - b^2$. Let $r_1 = PF_1$ and $r_2 = PF_2$. Then $r_1 + r_2 = 2a$ and

$$r_1r_2 = (\tfrac{1}{2})[(r_1 + r_2)^2 - r_1^2 - r_2^2]$$
$$= (\tfrac{1}{2})[4a^2 - (x + c)^2 - y^2 - (x - c)^2 - y^2]$$
$$= 2a^2 - x^2 - y^2 - c^2 = a^2 + b^2 - x^2 - y^2.$$

A point (u, v) on the tangent to the ellipse at P satisfies

$$\frac{xu}{a^2} + \frac{yv}{b^2} = 1.$$

Putting this in the form $u \cos\theta + v \sin\theta = d$, one finds that

$$d^2 = \frac{1}{(x/a^2)^2 + (y/b^2)^2} = \frac{a^4 b^4}{b^4 x^2 + a^4 y^2}.$$

But $b^4x^2 + a^4y^2 = b^2(a^2b^2 - a^2y^2) + a^2(a^2b^2 - b^2x^2) = a^2b^2(a^2 + b^2 - x^2 - y^2) = a^2b^2r_1r_2$. Hence $d^2r_1r_2 = a^4b^4r_1r_2/a^2b^2r_1r_2 = a^2b^2$, a constant.

B–5.

The sum is $n!$ since it is an nth difference of a monic polynomial, x^n, of degree n.

B–6.

Let $N = 2^\alpha p_1^{\beta_1} p_2^{\beta_2} \cdots p_k^{\beta_k}$ where α and the β_i are nonnegative integers and the p_i are distinct odd primes. Then

$$\sigma(N) = \sigma(2^\alpha)\sigma(p_1^{\beta_1}) \cdots \sigma(p_k^{\beta_k}).$$

Since $\sigma(N) = 2N + 1$ is odd, it follows that $\sigma(p_i^{\beta_i})$ is odd, $1 \leqq i \leqq k$. But

$$\sigma(p_i^{\beta_i}) = 1 + p_i + p_i^2 + \cdots + p_i^{\beta_i}$$

is odd if and only if β_i is even; for if β_i were odd, the right hand side would be the sum of an even number of odd numbers and hence even. It follows that the odd part of N must be a square, so that we may write

$$N = 2^\alpha M^2,\ \alpha \geqq 0. \tag{1}$$

where M is odd. It remains to show that $\alpha = 0$.

Since N is quasiperfect, $\sigma(N) = 2^{\alpha+1}M^2 + 1$, while from (1) we deduce $\sigma(N) = \sigma(2^\alpha)\sigma(M^2) = (2^{\alpha+1} - 1)\sigma(M^2)$. Hence $2^{\alpha+1}M^2 + 1 = (2^{\alpha+1} - 1)\sigma(M^2)$ so that

$$M^2 + 1 \equiv 0 \pmod{2^{\alpha+1} - 1}. \tag{2}$$

If $\alpha > 0$, $2^{\alpha+1} - 1 \equiv 3 \pmod 4$. Consequently $2^{\alpha+1} - 1$ has a prime divisor $p \equiv 3 \pmod 4$. Equation (2) implies

$$M^2 + 1 \equiv 0 \pmod p. \tag{3}$$

But since -1 is a quadratic non-residue modulo p whenever $p \equiv 3 \pmod 4$, (3) is impossible. Thus, $\alpha = 0$.

THE THIRTY-EIGHTH WILLIAM LOWELL PUTNAM MATHEMATICAL COMPETITION

December 3, 1977

A-1.

A line meeting the graph in four points has an equation $y=mx+b$. Then the x_i are the roots of

$$2x^4+7x^3+(3-m)x-(5+b)=0,$$

their sum is $-7/2$, and their arithmetic mean $(\Sigma x_i)/4$ is $-7/8$, which is independent of the line.

A-2.

We show that w must equal one of x, y, z and that the remaining two unknowns must be negatives of each other. Let $s=x+y$ and $p=xy$. Then the given equations imply that $w-z=s$ and that

$$\frac{s}{p}=\frac{x+y}{xy}=\frac{1}{y}+\frac{1}{x}=\frac{1}{w}-\frac{1}{z}=\frac{z-w}{zw}=-\frac{s}{zw}.$$

Then $s/p=s/(-zw)$ implies that either $s=0$ or $-zw=p$. If $s=0$, then $y=-x$ and $w=z$. If $-zw=p=xy$, then $-z$ and w are the roots of the quadratic equation $T^2-sT+p=0$, which has x and y as its roots; this case thus leads to either $w=x$ and $-z=y$ or $w=y$ and $-z=x$.

A-2. *Alternate solution*

Substitute w from the first equation into the second. The resulting expression factors into $(x+y)(x+z)(y+z)=0$. In this form we see that two of the numbers x, y, z must be negatives of each other, and the third number must equal w (by the first equation).

A−3.

We show that there are an infinite number of expressions for $u(x)$ in terms of f and g; some of the simpler ones are:

$$\begin{aligned} u(x)&=g(x)-f(x+3)+f(x+1)+f(x-1)-f(x-3)\\ &=-g(x+2)+f(x+5)-f(x+3)+f(x+1)+f(x-1)\\ &=g(x+4)-f(x+7)+f(x+5)-f(x+3)+f(x+1). \end{aligned}$$

Let E be the shift operator on functions A defined by $EA(x)=A(x+1)$. Then $(E+E^{-1})u(x)=2f(x)$ and $(E^4+E^{-4})u(x)=2g(x)$ are given. Thus $(E^2+1)u(x)=2Ef(x)$ and $(E^8+1)u(x)=2E^4g(x)$. Motivated by the fact that E^2+1 and E^8+1 are relatively prime polynomials in E, one finds that

$$1=\frac{1}{2}(E^8+1)-\frac{1}{2}(E^6-E^4+E^2-1)(E^2+1),$$

$$u(x)=\frac{1}{2}(E^8+1)u(x)-\frac{1}{2}(E^6-E^4+E^2-1)(E^2+1)u(x),$$

$$u(x)=E^4g(x)-(E^6-E^4+E^2-1)Ef(x),$$

$$u(x)=E^4g(x)+(-E^7+E^5-E^3+E)f(x),$$

$$u(x)=g(x+4)-f(x+7)+f(x+5)-f(x+3)+f(x+1).$$

Other expressions are obtained using

$$g(y)=-g(y-2)+f(y+3)+f(y-5)$$

$$=-g(y+2)+f(y+5)+f(y-3).$$

A-4.

$$\sum_{n=0}^{N}\frac{x^{2^n}}{1-x^{2^{n+1}}}=\sum_{n=0}^{N}\left(\frac{1}{1-x^{2^n}}-\frac{1}{1-x^{2^{n+1}}}\right)$$

$$=\frac{1}{1-x}-\frac{1}{1-x^{2^{N+1}}}\to\frac{1}{1-x}-1=\frac{x}{1-x}\text{ as } N\to\infty,$$

since $|x|<1$.

A-5.

It is well known that $\binom{p}{i}\equiv 0 \bmod p$ for $i=1,2,\cdots,p-1$ or equivalently that in $Z_p[x]$ one has $(1+x)^p=1+x^p$, where Z_p is the field of the integers modulo p. Thus in $Z_p[x]$,

$$\sum_{k=0}^{pa}\binom{pa}{k}x^k=(1+x)^{pa}=\left[(1+x)^p\right]^a=[1+x^p]^a=\sum_{j=0}^{a}\binom{a}{j}x^{jp}.$$

Since coefficients of like powers must be congruent modulo p in the equality

$$\sum_{k=0}^{pa}\binom{pa}{k}x^k=\sum_{j=0}^{a}\binom{a}{j}x^{jp}$$

in $Z_p[x]$, one sees that

$$\binom{pa}{pb}\equiv\binom{a}{b}(\bmod p)$$

for $b=0,1,\ldots,a$.

A-6.

For (a,b) in S, let $I(a,b)$ be $\iint f(x,y)\,dx\,dy$ over the rectangle $0\leqslant x\leqslant a$, $0\leqslant y\leqslant b$. Also let (a,b) define inductively a sequence (a_n,b_n) using $a_1=a, b_1=b$, $a_{n+1}=a_n-b_n$ and $b_{n+1}=b_n$ when $0\leqslant b_n\leqslant a_n$, and $a_{n+1}=a_n$ and $b_{n+1}=b_n-a_n$ when $0\leqslant a_n<b_n$. Then the hypothesis implies that $I(a,b)=I(a_n,b_n)$ for all n. Since f is bounded on S and $\lim_{n\to\infty}a_n=0=\lim_{n\to\infty}b_n$, it follows that $I(a,b)=0$ for all (a,b) in S.

If $f(x,y)$ is not zero for all (x,y) in S, then f must be positive (or negative) in some rectangle $R=\{(x,y):c\leqslant x\leqslant d, h\leqslant y\leqslant k\}$ and hence $I=\int_R\int f(x,y)dxdy$ must be positive (or negative). But this contradicts

$$I=I(h,k)-I(h,d)-I(c,k)+I(c,d)=0.$$

Thus f is identically zero on S.

B-1.

$$\prod_{n=2}^{\infty}\frac{n^3-1}{n^3+1}=\lim_{k\to\infty}\left[\frac{2^3-1}{2^3+1}\cdot\frac{3^3-1}{3^3+1}\cdots\frac{k^3-1}{k^3+1}\right]$$

$$= \lim_{k\to\infty}\left[\frac{1\cdot 7}{3\cdot 3}\cdot\frac{2\cdot 13}{4\cdot 7}\cdot\frac{3\cdot 21}{5\cdot 13}\cdots\frac{(k-1)(k^2+k+1)}{(k+1)(k^2-k+1)}\right]$$

$$= \lim_{k\to\infty}\left[\frac{2}{3}\cdot\frac{k^2+k+1}{k(k+1)}\right]=\frac{2}{3}.$$

B-2.

Let O' be any point different from O on the line of intersection of planes AOC and BOD, e.g., O' may be the intersection of lines AC and BD. Let A' be the intersection of line OA with the line through O' and parallel to OC. Let C' be the intersection of line OC with the line through O' which is parallel to OA. Then $OA'O'C'$ is a parallelogram and its diagonals OO' and $A'C'$ bisect each other at a point M. Choosing B' and D' in the same way, one obtains a parallelogram $OB'O'D'$ whose diagonals OO' and $B'D'$ also bisect each other at the midpoint M of segment OO'. Hence segments $A'C'$ and $B'D'$ bisect each other (at M) and $A'B'C'D'$ is a parallelogram. (The parallelogram is not unique.)

B-3.

Let $x_i=\sum_{j=1}^{\infty}a_{ij}2^{-j}$, with $a_{ij}\in\{0,1\}$, be the binary expansion of x_i. The triple is balanced if $a_{11}=a_{21}=a_{31}=0$. Otherwise, $a_{i1}=1$ for exactly one i and the balancing act produces $x_i'=\sum_{j=1}^{\infty}a_{i,j+1}2^{-j}$. An unbalanced triple that remains unbalanced after any finite number of balancing acts is constructed by choosing the a_{ij} so that exactly one of a_{1j}, a_{2j}, a_{3j} equals 1 for each j while taking care that no one of sequences $a_{i1}, a_{i2},\ldots$ repeats in blocks, i.e., that each x_i is irrational. One such solution has

$$\begin{aligned} a_{1j}&=1 \quad \text{if and only if } j\in\{1,9,25,49,\ldots\},\\ a_{2j}&=1 \quad \text{if and only if } j\in\{4,16,36,64,\ldots\},\\ a_{3j}&=1 \quad \text{if and only if } j\in\{2,3,5,6,\ldots\}.\end{aligned}$$

B-4.

We can assume that $Q=O$, the origin. Let $-C$ be the image of C under the reflection $P\to -P$. $-C$ is again a continuous closed curve surrounding O and $C\cap -C\neq\varnothing$ since they have the same diameter and both surround O (hence neither can be exterior to the other). Let $P_1\in C\cap -C$. Then there exists $P_2\in C$ such that $P_1=-P_2$. These are the two desired points.

B-5.

From the Cauchy–Schwarz Inequality, one has

$$[(a_1+a_2)+a_3+a_4+\cdots+a_n]^2\leqslant[1^2+1^2+\cdots+1^2][(a_1+a_2)^2+a_3^2+\cdots+a_n^2]$$

or

$$(\Sigma a_i)^2\leqslant(n-1)[(\Sigma a_i^2)+2a_1a_2] \quad\text{or}\quad [1/(n-1)](\Sigma a_i)^2\leqslant(\Sigma a_i^2)+2a_1a_2.$$

Using the hypothesis, one then has

$$A<-(\Sigma a_i^2)+\frac{1}{n-1}(\Sigma a_i)^2\leqslant-(\Sigma a_i^2)+(\Sigma a_i^2)+2a_1a_2=2a_1a_2.$$

Similarly, $A<2a_ia_j$ for $1\leqslant i<j\leqslant n$.

B-6.

Clearly $1\in H$. Also $x\in H$ implies $x^{-1}\in H$. Then the hypothesis implies that $a^{-1}=a^2$ and that $xaxaxa=1=x^{-1}ax^{-1}ax^{-1}a$ when $x\in H$. Thus one easily shows that

$$\text{(i) } axa=x^{-1}a^2x^{-1}, \qquad \text{(ii) } a^2xa^2=x^{-1}ax^{-1}.$$

Let

$$A=\{xay:x,y\in H\},\qquad B=\{xa^2y:x,y\in H\},$$
$$C=\{xa^2ya:x,y\in H\},\quad \text{and}\quad Q=A\cup B\cup C.$$

Each of A, B, C has at most h^2 elements; hence Q has at most $3h^2$ elements. Thus it suffices to prove that $x_1ax_2a\cdots x_na\in Q$ when each $x_i\in H$. We do this by induction on n.

For $n=1$, one sees that $x_1a=x_1a\cdot 1\in A\subseteq Q$. Now let $x_1,x_2,\ldots,x_{k+1}\in H$; $x_1ax_2a\cdots x_ka=q, x_{k+1}=z$, and $qza=p$. Inductively, we assume $q\in Q$ and seek to show $p\in Q$. The assumption implies that q is in A, B, or C. If $q=xay\in A$, then

$$p=(xay)za=xa(yz)a=x(yz)^{-1}a^2(yz)^{-1}\in B\subseteq Q,$$

using (i). If $q=xa^2y\in B$, then $p=xa^2yza\in C\subseteq Q$. If $q=xa^2ya\in C$, then

$$p=xa^2y(aza)=xa^2y(z^{-1}a^2z^{-1})=x[a^2(yz^{-1})a^2]z^{-1}$$
$$=x(yz^{-1})^{-1}a(yz^{-1})^{-1}z^{-1}\in A\subseteq Q,$$

using (i) and (ii).

THE THIRTY-NINTH WILLIAM LOWELL PUTNAM MATHEMATICAL COMPETITION

December 2, 1978

A-1.

Each of the twenty integers of A must be in one of the eighteen disjoint sets

$$\{1\}, \{52\}, \{4,100\}, \{7,97\}, \{10,94\}, \ldots, \{49,55\}.$$

Hence some (at least two) of the pairs $\{4,100\}, \ldots, \{49,55\}$ must have two integers from A. But the sum for each of these pairs is 104.

A-2.

Let M_t be the matrix obtained by subtracting t from each entry of the given matrix and let $G(t)$ be the determinant of M_t. By subtracting the entries of any row from the corresponding entries of each other row, one sees that $G(t)$ is linear in t. Then one notes that $G(a)=f(a)$ and $G(b)=f(b)$ using the fact that they are determinants of triangular matrices. Then linear interpolation shows that the desired determinant $G(0)$ is

$$[bG(a)-aG(b)]/(b-a)=[bf(a)-af(b)]/(b-a).$$

A-2. *Alternate solution*

A solution can be given using induction. For the inductive step, subtract the second column from the first and expand by cofactors down the first column. Apply the inductive assumption to the two resulting determinants using the functions $F(x)=(p_2-x)(p_3-x)\cdots(p_k-x)$ and $G(x)=(a-x)(p_3-x)\cdots(p_k-x)$ respectively. Simplify the resulting expression, noting that $G(a)=0$ and $(p_1-a)F(a)=f(a)$.

A-3.

Since the integral converges for $-1<k<5$, one can consider I_k to be defined on this open interval. Letting $x=1/t$, one finds that

$$I_k=\int_\infty^0 \frac{t^{-k}}{t^{-6}p(t)}\left(\frac{-dt}{t^2}\right)=\int_0^\infty \frac{t^{4-k}dt}{p(t)}=I_{4-k}.$$

Then

$$I_k=(I_k+I_{4-k})/2=\int_0^\infty \frac{[(x^k+x^{4-k})/2]dx}{p(x)} \geqslant \int_0^\infty \frac{x^2dx}{p(x)}=I_2,$$

since $(x^k+x^{4-k})/2 \geqslant \sqrt{x^k \cdot x^{4-k}}=x^2$ by the Arithmetic Mean–Geometric Mean Inequality. Thus I_k is smallest for $k=2$.

A-4.

(a) The defining property with $[w,x,y,z]=[a,b,a,b]$ and the hypothesis $B(a,b)=c$ give us

$$B(c,c)=B(B(a,b),B(a,b))=B(a,b)=c.$$

(b) The defining property with $[w,x,y,z]=[a,b,x,x]$ and $B(a,b)=c$ give

$$B(c,B(x,x))=B(B(a,b),B(x,x))=B(a,x).$$

Then using the result in (a) and $[w,x,y,z]=[c,c,x,x]$, one has

$$B(c,B(x,x))=B(B(c,c),B(x,x))=B(c,x).$$

Together, these show that $B(a,b)=c$ implies $B(a,x)=B(c,x)$ for all x in S.

(c) An easy way to obtain a bypass with property (i) is to let S be a cartesian product $I\times J$ and to define the operation B by

$$B((i,j),(h,k))=(i,k).$$

Properties (ii) and (iii) will hold if I and J, respectively, have more than one element. Except for notation, every bypass is obtained this way.

Tables with $S=\{a,b,c,d\}$ and with $S=\{u,v,w,x,y,z\}$ follow:

	a or c	b or d
a or b	a	b
c or d	c	d

	u or x	v or y	w or z
u or v or w	u	v	w
x or y or z	x	y	z

A-5.

Let $g(x)=\ln[(\sin x)/x]=\ln(\sin x)-\ln x$. Then

$$g''(x)=-\csc^2 x+\frac{1}{x^2}=\frac{1}{x^2}-\frac{1}{\sin^2 x}<0 \quad \text{for } 0<x<\pi$$

since $x>\sin x$ for $x>0$. Thus the graph of $g(x)$ is concave down and hence

$$\frac{1}{n}\sum_{i=1}^{n} g(x_i)\leqslant g\left(\frac{\sum_{i=1}^{n} x_i}{n}\right)=g(x),$$

or $\Sigma\, g(x_i)\leqslant ng(x)$. Since e^x is an increasing function, this implies

$$\prod_{i=1}^{n}\frac{\sin x_i}{x_i}=e^{\Sigma g(x_i)}\leqslant e^{ng(x)}=\left(\frac{\sin x}{x}\right)^n.$$

A-6.

For a set $\{p_1,\ldots,p_n\}$ of points in the plane, let e_i be the number of p_j one unit from p_i. Then $E=(e_1+\cdots+e_n)/2$ is the number of pairs with unit distance. Let C_i be the circle with center at p_i and radius 1. Each pair of circles has at most 2 intersections, so the C_i intersect in at most $2\binom{n}{2}=n(n-1)$ points. It suffices to treat the case in which each $e_i\geqslant 1$.

The point p_i occurs $\binom{e_i}{2}$ times as an intersection of C_j. Hence

(A) $$n(n-1)\geqslant\sum\binom{e_i}{2}=\sum e_i(e_i-1)/2\geqslant(1/2)\sum(e_i-1)^2.$$

In (A) and what follows, all sums are over $i=1,2,\ldots,n$. Using the Cauchy-Schwarz Inequality and (A) one has

$$\left[\sum(e_i-1)\right]^2 \leqslant \left[\sum 1\right]\left[\sum(e_i-1)^2\right] \leqslant n \cdot 2n(n-1) < 2n^3.$$

Hence $\Sigma(e_i-1) \leqslant \sqrt{2}\, n^{3/2}$ and so

$$E=\left(\sum e_i\right)/2 \leqslant (n+\sqrt{2}\, n^{3/2})/2 < 2n^{3/2}.$$

A-6. *Alternate solution*

Let $f(n)$ be the maximum number of pairs with unit distance apart from a set of n points in a plane. Then $f(1)=0$, $f(2)=1$, $f(3)=3$, and $f(4)=5$.

Suppose we have an array of $n \geqslant 2$ points which realizes $f(n)$ pairs unit distance apart. Suppose each point has at least k points (in the set) unit distance away and that one of them, namely x_0, has exactly k points unit distance away. Let these points be $x_1,\dots,x_k$.

For $i=0,1,\dots,k$, let C_i be the circle with radius 1 and center x_i. For $i>0$, C_i goes through x_0 and at most two other x_j. Therefore, there are at least $k-3$ points of the array other than $x_0,\dots,x_k$ on C_i. Any one of these $k-3$ points can appear on at most one other C_j since there are only two unit circles through x_0 and another point. Hence

$$n \geqslant 1+k+\frac{k(k-3)}{2}=1+\frac{k(k-1)}{2}.$$

Thus k is such that the triangular number $k(k-1)/2$ is less than n; we let the largest such integer k be k_n.

The array with x_0 removed has at most $f(n-1)$ pairs unit distance apart; therefore $f(n) \leqslant f(n-1)+k_n$. By repeating this argument we find that $f(n) \leqslant k_2+k_3+\cdots+k_n$.

The definition of k_n implies that $k_n=t$ for $\binom{t}{2} < n \leqslant \binom{t+1}{2}$. From this it follows that $k_n \leqslant 1+\sqrt{2(n-1)}$. Hence

$$\begin{aligned} f(n) &\leqslant k_2+k_3+\cdots+k_n \\ &\leqslant n-1+\sqrt{2}\,(\sqrt{1}+\sqrt{2}+\cdots+\sqrt{n-1}) \\ &\leqslant n-1+\sqrt{2}\int_1^n \sqrt{x}\,dx \\ &= n-1+2\sqrt{2}\,(n^{3/2}-1)/3 \\ &\leqslant n+.95n^{3/2} < 2n^{3/2}. \end{aligned}$$

B-1.

The area is the same as for an octagon inscribed in a circle and with sides alternately 3 units and 2 units in length. For such an octagon, all angles measure $3\pi/4$ and one can augment the octagon into a square with sides of length $3+2\sqrt{2}$ by properly placing a $\sqrt{2}$, $\sqrt{2}$, 2 isosceles right triangle on each of the sides of length 2. Hence the desired area is

$$(3+2\sqrt{2}\,)^2-(4\sqrt{2}\cdot\sqrt{2}\,/2)=13+12\sqrt{2}\,.$$

A second solution follows. Let r be the radius of the circle and let α and β be half of the central angles for the chords of lengths 3 and 2, respectively. Then $8\alpha+8\beta=2\pi$ and so $\beta=(\pi/4)-\alpha$. Also

$$\frac{3}{2r}=\sin\alpha, \qquad \frac{1}{r}=\sin\beta=\sin\left(\frac{\pi}{4}-\alpha\right)=\frac{\cos\alpha-\sin\alpha}{\sqrt{2}},$$

$$\frac{2}{3}=\frac{2r}{3}\cdot\frac{1}{r}=\frac{\cos\alpha-\sin\alpha}{\sqrt{2}\,\sin\alpha}=\frac{\cot\alpha-1}{\sqrt{2}}.$$

Now $\cot\alpha=(3+2\sqrt{2})/3=[(3+2\sqrt{2})/2]/(3/2)$ and hence the distance from the center of the circle to a chord of length 3 is $h_3=(3+2\sqrt{2})/2$. Similarly the distance to a chord of length 2 is $h_2=(2+3\sqrt{2})/2$. Finally, the desired area is

$$4(3h_3+2h_2)/2=(9+6\sqrt{2})+(4+6\sqrt{2})=13+12\sqrt{2}.$$

B-2.

Let S be the desired sum. Then

$$\begin{aligned} S&=\sum_{n=1}^{\infty}\frac{1}{n}\sum_{m=1}^{\infty}\frac{1}{n+2}\left(\frac{1}{m}-\frac{1}{m+n+2}\right)\\ &=\sum_{n=1}^{\infty}\frac{1}{n(n+2)}\left[\left(1-\frac{1}{n+3}\right)+\left(\frac{1}{2}-\frac{1}{n+4}\right)+\left(\frac{1}{3}-\frac{1}{n+5}\right)+\cdots\right]\\ &=\frac{1}{2}\sum_{n=1}^{\infty}\left(\frac{1}{n}-\frac{1}{n+2}\right)\left[\left(1-\frac{1}{n+3}\right)+\left(\frac{1}{2}-\frac{1}{n+4}\right)+\cdots\right].\end{aligned}$$

Hence

$$\begin{aligned} 2S&=\sum_{n=1}^{\infty}\left(\frac{1}{n}-\frac{1}{n+2}\right)\lim_{k\to\infty}\left[1+\frac{1}{2}+\cdots+\frac{1}{n+2}-\frac{1}{k}-\frac{1}{k+1}-\cdots-\frac{1}{k+n+1}\right]\\ &=\sum_{n=1}^{\infty}\left(\frac{1}{n}-\frac{1}{n+2}\right)\left(1+\frac{1}{2}+\cdots+\frac{1}{n+2}\right)\\ &=\lim_{h\to\infty}\left[\left(1-\frac{1}{3}\right)\left(1+\frac{1}{2}+\frac{1}{3}\right)+\left(\frac{1}{2}-\frac{1}{4}\right)\left(1+\frac{1}{2}+\frac{1}{3}+\frac{1}{4}\right)\right.\\ &\qquad\left.+\cdots+\left(\frac{1}{h}-\frac{1}{h+2}\right)\left(1+\frac{1}{2}+\cdots+\frac{1}{h}\right)\right]\\ &=\lim_{h\to\infty}\left[1\cdot\left(1+\frac{1}{2}+\frac{1}{3}\right)+\frac{1}{2}\left(1+\frac{1}{2}+\frac{1}{3}+\frac{1}{4}\right)+\frac{1}{3}\left(\frac{1}{4}+\frac{1}{5}\right)+\frac{1}{4}\left(\frac{1}{5}+\frac{1}{6}\right)\right.\\ &\quad\left.+\cdots+\frac{1}{h}\left(\frac{1}{h+1}+\frac{1}{h+2}\right)-\frac{1}{h+1}\left(1+\frac{1}{2}+\cdots+\frac{1}{h-1}\right)-\frac{1}{h+2}\left(1+\frac{1}{2}+\cdots+\frac{1}{h}\right)\right]\\ &=\frac{6+3+2}{6}+\frac{12+6+4+3}{2\cdot 12}+\left(\frac{1}{3\cdot 4}+\frac{1}{4\cdot 5}+\cdots\right)+\left(\frac{1}{3\cdot 5}+\frac{1}{4\cdot 6}+\cdots\right)\\ &=\frac{11}{6}+\frac{25}{24}+\frac{1}{3}+\frac{1}{2}\left(\frac{1}{3}+\frac{1}{4}\right)=\frac{7}{2}.\end{aligned}$$

Thus $S=7/4$.

B-3.

Clearly, $x_1=-1$, $x_2=-\frac{1}{2}$. An easy induction shows that each Q_n is positive for $x\geqslant 0$. Hence $x_n<0$, if Q_n has zeros.

Assume inductively that $x_1<x_2<\cdots<x_{2m-1}<x_{2m}$. Then $Q_{2m-1}(x)>0$ for $x>x_{2m-1}$. In particular, $Q_{2m-1}(x_{2m})>0$. Hence

$$\begin{aligned} Q_{2m+1}(x_{2m})&=Q_{2m}(x_{2m})+(m+1)x_{2m}Q_{2m-1}(x_{2m})\\ &=(m+1)x_{2m}Q_{2m-1}(x_{2m})<0.\end{aligned}$$

This implies that $Q_{2m+1}(x)=0$ for some $x>x_{2m}$, i.e., $x_{2m+1}>x_{2m}$. Similarly, one shows that $x_{2m+2}>x_{2m+1}$.

Let $a=-1/(m+1)$. Using the given recursive definition of the $Q_n(x)$, one finds that

$$Q_{2m+2}(a)=Q_{2m+1}(a)-Q_{2m}(a)=-Q_{2m-1}(a).$$

Hence at least one of $Q_{2m+2}(a)$ and $Q_{2m-1}(a)$ is nonpositive. Thus either $x_{2m+2}\geqslant a$ or $x_{2m-1}\geqslant a$. But each of these implies that both $x_{2m+2}\geqslant -1/(m+1)$ and $x_{2m+3}\geqslant -1/(m+1)$. It follows that $-2/n\leqslant x_n<0$ for all n and then that $\lim_{n\to\infty}x_n=0$.

B-4.

Clearly (1, 1, 1, 1) is a solution. Thinking of x_1, x_2, x_3 as fixed, the equation is quadratic in x_4 and one sees that the x_4 of a solution can be replaced by $x_4'=x_1x_2+x_1x_3+x_2x_3-x_4$ to obtain a new solution when $x_4'\neq x_4$. Also, the x_i may be permuted arbitrarily since the equation is symmetric in the x_i. Thus we may assume that $x_4\leqslant m=\min(x_1,x_2,x_3)$. Also assume that each $x_i\geqslant 1$. Then $x_4'\geqslant 3m^2-m>m$. This implies that one can start with the solution (1, 1, 1, 1) and through repeated use of the procedures stated above obtain a solution with each x_i an integer greater than N.

B-5.

The solution is very easy if one knows that the Chebyshev polynomial $C(x)=8x^4-8x^2+1=\cos(4\operatorname{Arccos}x)$ has the largest leading coefficient of all fourth degree polynomials $f(x)$ satisfying $-1\leqslant f(x)\leqslant 1$ for $-1\leqslant x\leqslant 1$; then one lets $P(x)=[C(x)+1]/2$ and has 4 as the largest A.

Without this information, one can use various substitutions to change the problem into equivalent ones of maximizing A in simpler functions satisfying conditions over intervals. With $Q(x)=[P(x)+P(-x)]/2$, the condition becomes

$$0\leqslant Q(x)=Ax^4+Cx^2+E\leqslant 1 \quad \text{over} \quad -1\leqslant x\leqslant 1.$$

Letting $x^2=y$, this becomes

$$0\leqslant R(y)=Ay^2+Cy+E\leqslant 1 \quad \text{over} \quad 0\leqslant y\leqslant 1.$$

Letting $y=(z+1)/2$ and $S(z)=R[(z+1)/2]$, one has

$$0\leqslant S(z)=(A/4)z^2+Fz+G\leqslant 1 \quad \text{over} \quad -1\leqslant z\leqslant 1.$$

With $T(z)=[S(z)+S(-z)]/2$, one obtains

$$0\leqslant (A/4)z^2+G\leqslant 1 \quad \text{over} \quad -1\leqslant z\leqslant 1.$$

Finally, letting $z^2=w$, it becomes

$$0\leqslant (A/4)W+G\leqslant 1 \quad \text{over} \quad 0\leqslant w\leqslant 1.$$

Now it is clear that the maximum A is 4 and that this maximum is achieved with $G=0$, i.e., with

$$T(z)=z^2, \qquad R(y)=(2y-1)^2, \qquad Q(x)=4x^4-4x^2+1.$$

B-6.

Let $a_h=(\Sigma_{k=1}^{ph}c_{h,k})/h$. Clearly, $0\leqslant a_h\leqslant p$. We now prove that $(\Sigma_{h=1}^{n}a_h)^2\leqslant 2p\Sigma_{h=1}^{n}(ha_h)$, which is equivalent to the assertion of the problem, by induction on n.

For $n=1$, one has $a_1^2\leqslant pa_1\leqslant 2pa_1$ as required. Suppose the inequality established for $n=m$. Then

$$\begin{aligned}\left(\sum_{h=1}^{m+1} a_h\right)^2 &= \left(\sum_{h=1}^{m} a_h\right)^2 + 2a_{m+1}\sum_{h=1}^{m} a_h + a_{m+1}^2 \\ &\leq 2p\sum_{h=1}^{m}(ha_h) + 2a_{m+1}pm + 2pa_{m+1} \\ &\leq 2p\left[(m+1)a_{m+1} + \sum_{h=1}^{m}(ha_h)\right] = 2p\sum_{h=1}^{m+1}(ha_h),\end{aligned}$$

as desired.

THE FORTIETH WILLIAM LOWELL PUTNAM MATHEMATICAL COMPETITION

December 1, 1979

A-1.

We see that $n=660$ and that all but one of the a_i equal 3 and the exceptional a_i is a 2 as follows. No a_i can be greater than 4 since one could increase the product by replacing 5 by $2\cdot 3$, 6 by $3\cdot 3$, 7 by $3\cdot 4$, etc. There cannot be both a 2 and a 4 or three 2's among the a_i since $2\cdot 4<3\cdot 3$ and $2\cdot 2\cdot 2<3\cdot 3$. Also, there cannot be two 4's since $4\cdot 4<2\cdot 3\cdot 3$. Clearly, no a_i is a 1. Hence the a_i are 3's except possibly for a 4 or for a 2 or for two 2's. Since $1979=3\cdot 659+2$, the only exception is a 2 and $n=660$.

A-2.

The condition is $k\geqslant 0$. If $k\geqslant 0$, one sees that $f(x)=\sqrt[4]{kx^3}$ satisfies $f(f(x))=kx^9$. For the converse, we note that $f(f(x))=kx^9$ for all real x with $k\neq 0$ implies that f takes on all real values since kx^9 does and implies that f is one-to-one since $f(a)=f(b)$ leads to $ka^9=f(f(a))=kb^9$ and hence $a=b$. But a *continuous* one-to-one function f from the real numbers $\mathbb{R}$ onto itself must be strictly monotonic. Also, if f is monotonic, either always increasing or always decreasing, $f(f(x))$ will always be increasing and so cannot equal kx^9 if $k<0$.

A-3.

The condition will be seen to be that $x_1=x_2=m$ for some integer m. Let $r_n=1/x_n$. Then $r_n=(2x_{n-2}-x_{n-1})/x_{n-2}x_{n-1}=2r_{n-1}-r_{n-2}$ and the r_n form an arithmetic progression. If x_n is a nonzero integer when n is in an infinite set S, the r_n for n in S satisfy $-1\leqslant r_n\leqslant 1$ and all but a finite number of the other r_n are also in this interval due to being nested among r_n with n in S; this can only happen if the r_n are all equal since the terms of an arithmetic progression are unbounded if the common difference $r_{n+1}-r_n$ is not zero. Equality of the r_n implies that $x_1=x_2=m$, an integer. Clearly, this condition is also sufficient.

Alternatively, let the r_n form the arithmetic progression defined above. If x_i and x_j are integers with $i\neq j$, then r_i and r_j and the common difference $(r_i-r_j)/(i-j)$ are rational. It follows that r_1 and r_2 are rational and hence that $r_1=a/q$ and $r_2=(a+d)/q$ with a, d, and q integers. Then $x_n=1/r_n=q/[a+(n-1)d]$. Since q has only a finite number of integral divisors, x_n can be an integer for an infinite set of n's only if $d=0$. This gives the same condition as in the first solution.

A-3. *Alternate solution*

An easy induction argument shows that

$$x_n=\frac{x_1x_2}{(n-1)x_1-(n-2)x_2}$$

$$=\frac{x_1x_2}{(x_1-x_2)n+(2x_2-x_1)},\qquad n=3,4,5,\ldots.$$

In this form we see that x_n will be an integer for infinitely many values of n if and only if $x_1 = x_2 = m$ for some integer m.

A-4.

There are a finite number (actually $n!$) of ways of pairing each of the red points with a blue point in a 1-to-1 way. Hence, there exists a pairing for which the sum of the lengths of the segments joining paired points is minimal. We now show that for such a pairing no two of the n segments intersect.

Let red points R and R' be paired with B and B', respectively, and assume that segments RB and $R'B'$ intersect. The triangle inequality implies that the sum of the lengths of these segments exceeds the sum of the lengths of segments RB' and $R'B$. Then interchanging B and B' would give us a new pairing with a smaller sum of segment lengths. This contradiction proves the existence of a pairing with nonintersecting segments.

A-5.

Let $f(x) = x^3 - 10x^2 + 29x - 25$. Then the table

x	1	2	3	5	6
$f(x)$	-5	1	-1	-5	5

shows that $f(x) = 0$ has three real solutions a, b, c with $1 < a < 2, 2 < b < 3, 5 < c < 6$. The number of integers that the set $\{1, 2, \ldots, n\}$ has in common with $S(a)$, $S(b)$, and $S(c)$ is $[n/a]$, $[n/b]$, and $[n/c]$, respectively. Since

$$\frac{1}{a} + \frac{1}{b} + \frac{1}{c} > \frac{1}{2} + \frac{1}{3} + \frac{1}{6} = 1,$$

one sees that

$$\lim_{n\to\infty} \left\{ \left[\frac{n}{a}\right] + \left[\frac{n}{b}\right] + \left[\frac{n}{c}\right] - n \right\} = \infty$$

and hence that an infinite number of positive integers appear in more than one of $S(a)$, $S(b)$, $S(c)$. This implies that some pair of these sets must have an infinite intersection.

A-6.

For $k = 0, 1, \ldots, 2n - 1$ let I_k be the open interval $(k/2n, [k+1]/2n)$. Among the $2n$ intervals I_k there exist n not containing any of the p_i and we place an x_j at the center of each of these n intervals. Let $|x_j - p_i| = d_{ij}$ and

$$B = 8n\left(1 + \frac{1}{3} + \frac{1}{5} + \cdots + \frac{1}{2n-1}\right).$$

For fixed i, the d_{ij} satisfy $d_{ij} \geqslant 1/4n$, at most two of them do not satisfy $d_{ij} \geqslant 3/4n$, at most four do not satisfy $d_{ij} \geqslant 5/4n$, etc. Hence

$$\sum_{j=1}^{n} \frac{1}{d_{ij}} \leqslant 2 \sum_{h=0}^{n-1} \frac{4n}{1+2h} = B.$$

(This inequality can be improved.) Thus we have

$$\sum_{j=1}^{n} \left(\sum_{i=1}^{n} \frac{1}{d_{ij}} \right) = \sum_{i=1}^{n} \left(\sum_{j=1}^{n} \frac{1}{d_{ij}} \right) \leqslant nB.$$

So clearly there is a value of j for which $\sum_{i=1}^{n}(1/d_{ij}) \leqslant B$ and the x_j for such a j can serve as the desired x.

B-1.

We assume that there is such a common normal and obtain a contradiction. This assumption implies

$$-\frac{a-c}{\cosh a-\sinh c}=\cosh c=\sinh a. \qquad \text{(I)}$$

Since $\cosh x>0$ for all real x and $\sinh x>0$ only for $x>0$, (I) implies $a>0$. Using the fact that $\sinh x<\cosh x$ for all x and (I), one obtains

$$\sinh c<\cosh c=\sinh a<\cosh a.$$

This, $a>0$, and the fact that $\cosh x$ increases for $x>0$ imply that $c<a$. Thus the leftmost expression in (I) is negative and cannot equal $\cosh c$. This contradiction shows that no common normal exists.

B-2.

Let $u=bx+a(1-x)$; then the definite integral becomes

$$I(t)=\frac{1}{b-a}\int_a^b u^t\,du=\frac{b^{t+1}-a^{t+1}}{(1+t)(b-a)}.$$

Using standard calculus methods for evaluating limits of indeterminate expressions, one finds that

$$[I(t)]^{1/t}\to e^{-1}(b^b/a^a)^{1/(b-a)} \text{ as } t\to 0.$$

B-3.

Let $r=(m-1)/2$. We show that $q(x)=p(x)+k$ is irreducible over F for r elements k of F. Since m is odd, the characteristic of F is not 2, $1+1=2\neq 0$, $2^{-1}b$ is an element h of F, the $2r+1$ elements of F can be expressed in the form $0,f_1,-f_1,\ldots,f_r,-f_r$, and $\{0,f_1^2,\ldots,f_r^2\}$ is the set of the $r+1$ distinct squares in F. Now

$$q(x)=(x+h)^2-(h^2-c-k)$$

is irreducible over F if and only if it has no zero in F, i.e., if and only if h^2-c-k is not one of the $r+1$ squares f^2 in F. Hence k must be one of the r elements left when the $r+1$ elements of the form h^2-c-f^2 are removed from the $2r+1$ elements of F.

B-3. *Alternate solution*

The polynomial x^2+bx+d is reducible if and only if there are elements s and t in F such that

$$st=d$$

$$s+t=-b,$$

or, equivalently, if and only if there is an element s in F such that $-s(s+b)=d$. For $s\in F$, define $f(s)=s(s+b)$. We have just seen that the number of reducible polynomials of the form x^2+bx+d is equal to the number of elements in the image of f. Notice that $f(s)=f(t)$ if and only if either $t=s$ or $t=-s-b$. Because F is not of characteristic 2, $s=-s-b$ for only one s in F. It follows that the image of F has $1+(m-1)/2$ distinct elements, and therefore the number of irreducible polynomials of the form $x^2+bx+c+k$ is $m-(1+(m-1)/2)=(m-1)/2$.

B-4.

(a) Trial of e^{mx} shows that $y=e^{3x}$ satisfies the homogeneous equation. Trial of a polynomial $x^d+\cdots$ shows that d must be 2 and then trial of x^2+px+q shows that $y=x^2+x$ is a solution. Any linear combination $he^{3x}+k(x^2+x)$, with at least one of the constants h and k not zero, is an answer.

(b) It is easy to see that $y=2$ satisfies the nonhomogeneous equation and hence $f(x)$ is of the form $2+he^{3x}+k(x^2+x)$. Now $f(0)=1$ gives us $2+h=1$ or $h=-1$. Then $[f(-1)-2][f(1)-6]=1$ leads to

$$-e^{-3}(2+2k-e^3-6)=1, \quad (2k-4)e^{-3}=0, \quad k=2.$$

Hence $f(x)=2-e^{3x}+2(x^2+x)$, $f(-2)=6-e^{-6}$, $f(2)=14-e^6$. Therefore we let $a=6$, $b=14$, and $c=1$.

We note that if one stops guessing after obtaining an answer $g(x)$ to (a), the standard substitutions $y=g(x)z$ followed by $z'=w$ will reduce the nonhomogeneous equation to a linear equation which can be solved by a well-known method.

B-5.

A support line for C is a straight line touching C such that one side of the line has no points of C. There is a support line containing $(0,1)$; let its slope be m. If $m \geqslant 1/2$, the part of the area of C in the fourth quadrant is no more than 1 and we are done. Similarly, if $m \leqslant -1/2$. So we assume that $-1/2<m<1/2$ and assume the analogous facts for support lines containing $(1,0)$, $(0,-1)$, and $(-1,0)$. At least one of the angles of the quadrilateral formed by these four support lines is not acute; we may take this angle α to be at a vertex (h,k) in the first quadrant. Then $\alpha \geqslant \pi/2$ implies that $h+k \leqslant 2$, and this in turn implies that the area of C in the first quadrant does not exceed 1. Hence $A(C) \leqslant 4$.

B-6.

Let $X=(x_1,\dots,x_n)$ and $Y=(y_1,\dots,y_n)$. Also let $a+bi$ be either square root of $z_1^2+\cdots+z_n^2$. Then $ab=X\cdot Y=x_1y_1+\cdots+x_ny_n$ and

$$a^2-b^2=\|X\|^2-\|Y\|^2=(x_1^2+\cdots+x_n^2)-(y_1^2+\cdots+y_n^2).$$

The Cauchy–Schwarz inequality tells us that $|X\cdot Y| \leqslant \|X\|\cdot\|Y\|$ and hence $|a|\cdot|b| \leqslant \|X\|\cdot\|Y\|$. Therefore, the assumption that $|a|>\|X\|$ would imply that $|b|<\|Y\|$. This and $a^2=\|X\|^2-\|Y\|^2+b^2$ would yield $a^2<\|X\|$ and thus the contradiction $|a|<\|X\|$. Hence the assumption is false and $r=|a| \leqslant \|X\|$. Since $\|X\|^2 \leqslant (|x_1|+\cdots+|x_n|)^2$, this implies the desired $r \leqslant |x_1|+\cdots+|x_n|$.

THE FORTY-FIRST WILLIAM LOWELL PUTNAM MATHEMATICAL COMPETITION

December 6, 1980

A-1.

We show that $g(x) = x^2 + bx + c - (1/4)$. The equation of the tangent to the given parabola at $P_j = (j, y_j)$ is easily seen to be $y = L_j$, where $L_j = (2j + b)x - j^2 + c$. Solving $y = L_j$ and $y = L_{j+1}$ simultaneously, one finds that $x = (2j + 1)/2$ and so $j = (2x - 1)/2$ at I_j. Substituting this expression for j into L_j gives the $g(x)$ above.

A-2.

We show that the number is $(1 + 4r + 6r^2)(1 + 4s + 6s^2)$. Each of a, b, c, d must be of the form $3^m 7^n$ with m in $\{0, 1, \dots, r\}$ and n in $\{0, 1, \dots, s\}$. Also m must be r for at least two of the four numbers, and n must be s for at least two of the four numbers. There is one way to have $m = r$ for all four numbers, $4r$ ways to have one m in $\{0, 1, \dots, r - 1\}$ and the other three equal to r, and $\binom{4}{2}r^2 = 6r^2$ ways to have two of the m's in $\{0, 1, \dots, r - 1\}$ and the other two equal to r. Thus there are $1 + 4r + 6r^2$ choices of allowable m's and, similarly, $1 + 4s + 6s^2$ choices of allowable n's.

A-3.

Let I be the given definite integral and $\sqrt{2} = r$. We show that $I = \pi/4$. Using $x = (\pi/2) - u$, one has

$$I = \int_{\pi/2}^{0} \frac{-du}{1 + \cot^r u} = \int_0^{\pi/2} \frac{\tan^r u \, du}{\tan^r u + 1}.$$

Hence

$$2I = \int_0^{\pi/2} \frac{1 + \tan^r x}{1 + \tan^r x} dx = \int_0^{\pi/2} dx = \pi/2 \quad \text{and} \quad I = \pi/4.$$

A-4.

(a). Let S be the set of the 10^{18} real numbers $r + s\sqrt{2} + t\sqrt{3}$ with each of r, s, t in $\{0, 1, \dots, 10^6 - 1\}$ and let $d = (1 + \sqrt{2} + \sqrt{3})10^6$. Then each x in S is in the interval $0 \leqslant x < d$. This interval is partitioned into $10^{18} - 1$ "small" intervals $(k - 1)e \leqslant x < ke$ with $e = d/(10^{18} - 1)$ and k taking on the values $1, 2, \dots, 10^{18} - 1$. By the pigeonhole principle, two of the 10^{18} numbers of S must be in the same small interval and their difference $a + b\sqrt{2} + c\sqrt{3}$ gives the desired a, b, c since $c < 10^{-11}$.

(b) Let $F_1 = a + b\sqrt{2} + c\sqrt{3}$ and F_2, F_3, F_4 be the other numbers of the form $a \pm b\sqrt{2} \pm c\sqrt{3}$. Using the irrationality of $\sqrt{2}$ and $\sqrt{3}$ and the fact that a, b, c are not all zero, one easily shows that no F_i is zero. (The demonstration of this was Problem A-1 of the 15th Competition, held on March 5, 1955. For the proof, see page 402 of *The William Lowell Putnam Mathematical Competition Problems and Solutions: 1938–1964*, published by the MAA, or see this MONTHLY, 62

(1955) 561.) One also sees readily that the product $P = F_1F_2F_3F_4$ is an integer. Hence $|P| \geq 1$. Then $|F_1| \geq 1/|F_2F_3F_4| > 10^{-21}$ since $|F_i| < 10^7$ and thus $1/|F_i| > 10^{-7}$ for each i.

A-5.

Let $Q = P - P'' + P^{iv} - \cdots$. Using repeated integrations by parts, the equations of the given system become

$$\int_0^x P(t)\sin t\,dt = -Q(x)\cos x + Q'(x)\sin x + Q(0) = 0,$$

$$\int_0^x P(t)\cos t\,dt = Q(x)\sin x + Q'(x)\cos x - Q'(0) = 0.$$

These imply that

$$Q(x) = Q'(0)\sin x + Q(0)\cos x. \qquad (E)$$

Since P and, hence, Q are polynomials of positive degree and the right side of (E) is bounded, equation (E) has all of its solutions in some interval $|x| \leq M$. In such an interval, $P(x)\sin x$ has only finitely many zeros and $\int_0^x P(t)\sin t\,dt = 0$ has at most one more zero by Rolle's Theorem. Q.E.D.

A-6.

We show that $u = 1/e$. Since $f' - f = (fe^{-x})'e^x$ and $e^x \geq 1$ for $x \geq 0$,

$$\int_0^1 |f' - f|\,dx = \int_0^1 |(fe^{-x})'e^x|\,dx \geq \int_0^1 (fe^{-x})'\,dx = [fe^{-x}]_0^1 = 1/e.$$

To see that $1/e$ is the largest lower bound, we use functions $f_a(x)$ defined by

$$f_a(x) = (e^{a-1}/a)x \quad \text{for} \quad 0 \leq x \leq a, \qquad f_a(x) = e^{x-1} \quad \text{for} \quad a \leq x \leq 1.$$

Let $m = e^{a-1}/a$. Then

$$\int_0^1 |f_a'(x) - f_a(x)|\,dx = \int_0^a |m - mx|\,dx = m\left(a - \frac{a^2}{2}\right) = e^{a-1}\left(1 - \frac{a}{2}\right).$$

As $a \to 0$, this expression approaches $1/e$. The function $f_a(x)$ does not have a continuous derivative, but one can smooth out the corner, keeping the change in the integral as small as one wishes, and thus show that no number greater than $1/e$ can be an upper bound.

B-1.

The inequality holds if and only if $c \geq 1/2$. For $c \geq 1/2$,

$$\frac{e^x + e^{-x}}{2} = \sum_{n=0}^{\infty} \frac{x^{2n}}{(2n)!} \leq \sum_{n=0}^{\infty} \frac{x^{2n}}{2^n n!} = e^{x^2/2} \leq e^{cx^2}$$

for all x since $(2n)! \geq 2^n n!$ for $n = 0, 1, \ldots$.

Conversely, if the inequality holds for all x, then

$$0 \leq \lim_{x\to 0} \frac{e^{cx^2} - \frac{1}{2}(e^x + e^{-x})}{x^2} = \lim_{x \to 0} \frac{(1 + cx^2 + \cdots) - (1 + \frac{1}{2}x^2 + \cdots)}{x^2} = c - \frac{1}{2}$$

and so $c \geq 1/2$.

B-2.

(a) $v = 7$. The seven vertices are $V_0 = (0,0,0)$, $V_1 = (11,0,0)$, $V_2 = (0,9,0)$, $V_3 = (0,0,8)$, $V_4 = (0,3,8)$, $V_5 = (9,0,2)$, and $V_6 = (4,7,0)$.

(b) $e = 11$. The eleven edges are V_0V_1, V_0V_2, V_0V_3, V_1V_5, V_1V_6, V_2V_4, V_2V_6, V_3V_4, V_3V_5, V_4V_5, and V_4V_6.

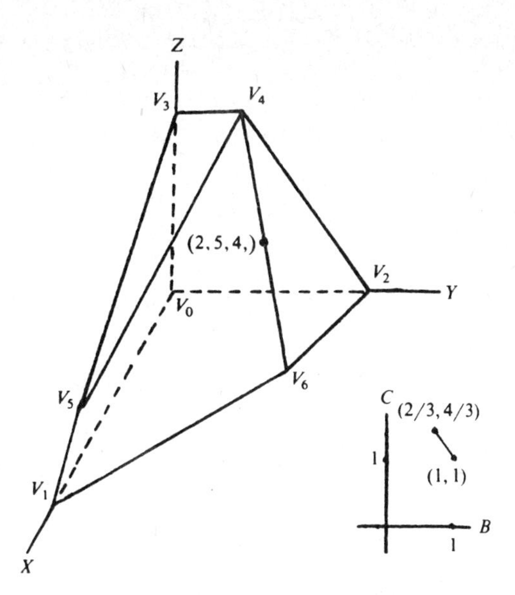

(c) The desired (b, c) are those with $b + c = 2$ and $2/3 \leqslant b \leqslant 1$. Let $L(x, y, z) = bx + cy + z$. Since L is linear and $(2, 5, 4)$ is on edge V_4V_6, the maximum of L on S must be assumed at V_4 and at V_6 and the conditions on b and c are obtained from $L(0, 3, 8) = L(4, 7, 0) \geqslant L(x, y, z)$, with (x, y, z) ranging over the other five vertices.

B-3.

We show that $u_n > 0$ for all $n \geqslant 0$ if and only if $a \geqslant 3$. Let $\Delta u_n = u_{n+1} - u_n$. Then the recursion (i.e., difference equation) takes the form $(1 - \Delta)u_n = n^2$. Since n^2 is a polynomial, a particular solution is

$$u_n = (1 - \Delta)^{-1} n^2 = (1 + \Delta + \Delta^2 + \cdots)n^2 = n^2 + (2n + 1) + 2 = n^2 + 2n + 3.$$

(This is easily verified by substitution.) The complete solution is $u_n = n^2 + 2n + 3 + k \cdot 2^n$, since $v_n = k \cdot 2^n$ is the solution of the associated homogeneous difference equation $v_{n+1} - 2v_n = 0$. The desired solution with $u_0 = a$ is $u_n = n^2 + 2n + 3 + (a - 3)2^n$. Since $\lim_{n \to \infty}[2^n/(n^2 + 2n + 3)] = +\infty$, u_n will be negative for large enough n if $a - 3 < 0$. Conversely, if $a - 3 \geqslant 0$, it is clear that each $u_n > 0$.

Alternatively, one sees that $u_0 = a$ and $u_1 = 2a$ and one can prove by mathematical induction that

$$u_n = 2^n a - \sum_{k=1}^{n-1} 2^{n-1-k} k^2 \quad \text{for} \quad n \geqslant 2.$$

Hence $u_n > 0$ for $n \geqslant 0$ if and only if $a > \sum_{k=1}^{n-1} 2^{-1-k} k^2$ and this holds if and only if $a \geqslant L$, where $L = \sum_{k=1}^{\infty} 2^{-1-k} k^2$. Let D mean d/dx. Then for $|x| < 1$,

$$(1 - x)^{-1} = \sum_{k=0}^{\infty} x^k$$

$$D(1 - x)^{-1} = (1 - x)^{-2} = \sum_{k=1}^{\infty} kx^{k-1}$$

$$D(1 - x)^{-2} = 2(1 - x)^{-3} = \sum_{k=2}^{\infty} k(k - 1)x^{k-2}.$$

Let $g(x) = 2x^3(1 - x)^{-3} + x^2(1 - x)^{-2}$. Then $L = g(1/2) = 3$ and the answer is all $a \geqslant 3$.

B-4.

The result we are asked to prove is clearly not true if $|X| < 10$. Hence we assume that $|X| \geqslant 10$ or that the A_j are distinct, which implies that $|X| \geqslant 10$.

Let $X = \{x_1, \ldots, x_m\}$, with $m = |X|$, and let n_i be the number of j such that x_i is in A_j. Let N be the number of ordered pairs (i, j) such that x_i is in A_j. Then

$$N = n_1 + n_2 + \cdots + n_m = |A_1| + |A_2| + \cdots + |A_{1066}| > 1066(m/2) = 533m.$$

Hence one of the n_i, say n_1, exceeds 533.

Let $B_1, \ldots, B_s$ be those sets A_j not containing x_1 and $Y = \{x_2, x_3, \ldots, x_m\}$. Then $s = 1066 - n_1 \leqslant 532$ and each $|B_j| > |Y|/2$. We can assume that x_2 is in at least as many B_j as any other x_i and let $C_1, \ldots, C_t$ be the B_j not containing x_2. As before, one can show that $t \leqslant 265$.

We continue in this way. The 4th sequence of sets $D_1, \ldots, D_u$ will number no more than 132. The numbers of sets in the 5th through 10th sequences will number no more than 65, 32, 15, 7, 3, and 1, respectively. Thus we obtain the desired elements $x_1, \ldots, x_{10}$ unless X has fewer than 10 elements.

B-5.

The answer is $1 \geqslant t$ (or $0 \leqslant t \leqslant 1$). The product fg of two nonnegative increasing continuous real-valued functions has the same properties. Using the fact that $0 \leqslant a \leqslant c$ and $0 \leqslant b \leqslant d$ imply $ad + bc \leqslant cb + cd$, one shows that fg is convex when f and g are convex. The function $f(x) = x$ is in S_t for all t. If S_t is closed under multiplication, x^2 is in S_t and so $2/9 = 1 - 2(4/9) + (1/9) \geqslant t[4/9 - 2(1/9)] = 2t/9$ or $1 \geqslant t$.

The following argument proves the converse. Let $t \in [0,1]$. For a real valued function h defined on $[0,1]$, let $E(h) = [h(1) - 2h(2/3) + h(1/3)] - t[h(2/3) - 2h(1/3) + h(0)]$. Suppose that f and g are in S_t, so $E(f) \geqslant 0$ and $E(g) \geqslant 0$. Then $E(fg) = g(2/3)E(f) + f(1/3)E(g) + [f(1) - f(1/3)][g(1) - g(2/3)] - t[f(1/3) - f(0)][g(2/3) - g(0)]$. By convexity, $f(1) - f(1/3) \geqslant 2[f(1/3) - f(0)]$, and $g(1) - g(2/3) \geqslant \frac{1}{2}[g(2/3) - g(0)]$. If $t \leqslant 1$, this implies $E(fg) \geqslant 0$, so fg is in S_t.

B-6.

Let $F_d(x) = \sum_{n=d}^{\infty} G(d, n)x^n$. Then $F_1(x) = \sum_{n=1}^{\infty} x^n/n$ and $F_1'(x) = \sum_{n=0}^{\infty} x^n$. One sees that $F_d'(x) = dF_{d-1}(x)F_1'(x)$ by finding the coefficients of x^{n-1} on both sides and using $nG(d, n) = d\sum_{i=d}^{n} G(d-1, i-1)$. Then an induction gives us $F_d(x) = [F_1(x)]^d$. Now, for $1 < d \leqslant p$, the coefficient $G(d, p)$ of x^p in $F_d(x)$ is the coefficient of x^p in $[\sum_{n=1}^{p-d+1} x^n/n]^d$, and hence $G(d, p) = s/t$ with s and t integers and t a product of primes less than p.

THE FORTY-SECOND WILLIAM LOWELL PUTNAM MATHEMATICAL COMPETITION

December 5, 1981

A-1.

We show that the limit is $1/8$. Let $T(m) = 1 + 2 + \cdots + m = m(m+1)/2$, $[x]$ denote the greatest integer in x, $h = [\log_5 n]$, and e_i be the fractional part $(n/5^i) - [n/5^i]$ for $1 \le i \le h$. Then

$$E(n) = 5T([n/5]) + 5^2T([n/5^2]) + \cdots + 5^hT([n/5^h])$$

$$2E(n) = 5([n/5]^2 + [n/5]) + 5^2([n/5^2]^2 + [n/5^2]) + \cdots + 5^h([n/5^h]^2 + [n/5^h])$$

$$= 5\left(\frac{n^2}{5^2} - \frac{2e_1n}{5} + e_1^2 + \frac{n}{5} - e_1\right) + \cdots + 5^h\left(\frac{n^2}{5^{2h}} - \frac{2e_hn}{5^h} + e_h^2 + \frac{n}{5^h} - e_h\right)$$

$$\frac{E(n)}{n^2} = \frac{1}{2}\left(\frac{1}{5} + \frac{1}{5^2} + \cdots + \frac{1}{5^h}\right) + \frac{h}{2n} - \frac{e_1 + e_2 + \cdots + e_h}{n}$$

$$+ \frac{5(e_1^2 - e_1) + \cdots + 5^h(e_h^2 - e_h)}{2n^2}.$$

Since $5^h \le n < 5^{h+1}$ and $0 \le e_i < 1$, one sees that $h/n \to 0$ and $E(n)/n^2 \to 1/8$ as $n \to \infty$.

A-2.

For any numbering, one can go from the square numbered 1 to the square numbered 64 in 7 or fewer steps, in each step going to an adjacent square; thus $(64 - 1)/7 = 9$ is a C-gap. It is the largest C-gap since with coordinates (a, b), $1 \le a \le 8$ and $1 \le b \le 8$, for the squares we can number (a, b) with $8(a - 1) + b$ and thus find that no number greater than 9 is a C-gap.

A-3.

Let $G(t)$ be the double integral. Then

$$\lim_{t \to \infty} [G(t)/e^t] = \lim_{t \to \infty} [G'(t)/e^t]$$

by L'Hôpital's Rule. One finds that

$$G'(t) = \int_0^t \frac{e^x - e^t}{x - t}\,dx + \int_0^t \frac{e^y - e^t}{y - t}\,dy = 2\int_0^t \frac{e^x - e^t}{x - t}\,dx.$$

Then using $e^x = e^t[1 + (x - t) + (x - t)^2/2! + \cdots]$, one sees that $e^{-t}G'(t) \to \infty$ as $t \to \infty$ since for sufficiently large t,

$$\frac{G'(t)}{2e^t} = \int_0^t \frac{e^{x-t} - 1}{x - t}\,dx = \int_0^t \frac{1 - e^{-y}}{y}\,dy > \int_1^t \frac{1 - e^{-y}}{y}\,dy > (1 - e^{-1})\log t.$$

A-4.

Set up coordinates so that a vertex of the given unit square is $(0,0)$ and two sides of the square are on the axes. Using the reflection properties, one can see that P escapes within T units of time if and only if the (infinite) ray from P_0, with the direction of the first segment of the path, goes through a lattice point (point with integer coordinates) within T units of distance from P_0. Thus $N(T)$ is at most the number $L(T)$ of lattice points in the circle with center at P_0 and radius T. Tiling the plane with unit squares having centers at the lattice points and considering areas, one sees that

$$N(T) \le L(T) \le \pi\left[T + (\sqrt{2}/2)\right]^2.$$

Hence there is an upper bound for $N(T)$ of the form $\pi T^2 + bT + c$, with b and c fixed. When just one coordinate of P_0 is irrational,

$$N(T) = L(T) \ge \pi\left[T - (\sqrt{2}/2)\right]^2.$$

This lower bound for $N(T)$ exceeds $aT^2 + bT + c$ for sufficiently large T if $a < \pi$; hence π is the desired a.

A-5.

We show that $Q(x)$ has at least $2n - 1$ real zeros. One finds that $Q(x) = F(x)G(x)$, where

$$F(x) = P'(x) + xP(x) = e^{-x^2/2}\left[e^{x^2/2}P(x)\right]',\, G(x) = xP'(x) + P(x) = [xP(x)]'.$$

We can assume that $P(x)$ has exactly n zeros a_i exceeding 1 with $1 < a_1 < a_2 < \cdots < a_n$. It follows from Rolle's Theorem that $F(x)$ has $n-1$ zeros b_i and $G(x)$ has n zeros c_i with

$$1 < a_1 < b_1 < a_2 < b_2 < \cdots < b_{n-1} < a_n, \quad 0 < c_1 < a_1 < c_2 < a_2 < \cdots < c_n < a_n.$$

If $b_i \neq c_{i+1}$ for all i, the b's and c's are $2n - 1$ distinct zeros of $Q(x)$. So we assume that $b_i = c_{i+1} = r$ for some i. Then

$$P'(r) + rP(r) = 0 = rP'(r) + P(r)$$

and so $(r^2 - 1)P(r) = 0$. Since $r = b_i > 1$, $P(r) = 0$. Since $a_i < r < a_{i+1}$, this contradicts the fact that the a's are all the zeros exceeding 1 of $P(x)$. Hence $Q(x)$ has at least $2n - 1$ distinct real zeros.

A-6.

Treating each point X of the plane as the vector $\overrightarrow{AX}$ with initial point at A and final point at X, let

$$L = (B + C)/2,\, M = C/2, \quad \text{and} \quad N = B/2$$

(be the midpoints of sides BC, AC, and AB). Also let

$$S = (2L + M)/3 = (B + C + M)/3,\, T = (2L + N)/3$$
$$= (B + C + N)/3,\, Q = 2P - B, \quad \text{and} \quad R = 3P - B - C.$$

Clearly Q and R are lattice points. Also $Q \neq P$ and $R \neq P$ since $Q = P$ implies $P = B$ and $R = P$ implies that P is the point L on side BC. Hence Q is not inside ΔABC and this implies that P is not inside ΔNBL since the linear transformation f with $f(X) = 2X - B$ translates a doubled ΔNBL (and its inside) onto ΔABC (and its inside). Similarly, P is not inside ΔMCL. Using the mapping $g(X) = 3X - B - C$ and the fact that R is not inside ΔLMN, one finds that P is not inside ΔLST. Since the distance from A to line ST is 5 times the distance between lines ST and BC, it follows that $|AP|/|PE| \le 5$. This upper bound 5 is seen to be the maximum by considering the example with $A = (0,0)$, $B = (0,2)$, and $C = (3,0)$ in which $P = (1,1) = T$ is the only lattice point inside ΔABC and $|AT|/|TE| = 5$.

B-1.

Let $S_k(n) = 1^k + 2^k + \cdots + n^k$. Using standard methods of calculus texts one finds that

$$S_2(n) = (n^3/3) + (n^2/2) + an$$

and

$$S_4(n) = (n^5/5) + (n^4/2) + bn^3 + cn^2 + dn,$$

with a, b, c, d constants. Then the double sum is

$$10nS_4(n) - 18[S_2(n)]^2 = (2n^6 + 5n^5 + \cdots) - (2n^6 + 6n^5 + \cdots) = -n^5 + \cdots$$

and the desired limit is -1.

B-2.

First we let $0 < a < b$ and seek the x that minimizes

$$f(x) = \left(\frac{x}{a} - 1\right)^2 + \left(\frac{b}{x} - 1\right)^2 \text{ on } a \le x \le b.$$

Let $x/a = z$ and $b/a = c$. Then

$$f(x) = g(z) = (z-1)^2 + \left(\frac{c}{z} - 1\right)^2.$$

Now $g'(z) = 0$ implies

$$z^4 - z^3 + cz - c^2 = (z^2 - c)(z^2 - z + c) = 0;$$

the only positive solution is $z = \sqrt{c}$. Since $0 < a < b, c > 1, \sqrt{c} > 1$, and

$$g(1) = g(c) = (c-1)^2 = (\sqrt{c} - 1)^2(\sqrt{c} + 1)^2 > 2(\sqrt{c} - 1)^2 = g(\sqrt{c}).$$

Hence the minimum of $g(z)$ on $1 \le z \le c$ occurs at $z = \sqrt{c}$. It follows that the minimum for $f(x)$ on $a \le x \le b$ occurs at $x = a\sqrt{b/a} = \sqrt{ab}$. Then the minimum for the given function of r, s, t occurs with $r = \sqrt{s}$, $t = \sqrt{4s} = 2r$, and $s = \sqrt{rt} = r\sqrt{2}$. These imply that $r = \sqrt{2}$, $s = 2$, $t = 2\sqrt{2}$. Thus the desired minimum value is $4(\sqrt{2} - 1)^2 = 12 - 8\sqrt{2}$.

B-3.

As m ranges through all nonnegative integers,

$$n = (m^2 + m + 2)(m^2 + m + 3) + 3$$

takes on an infinite set of positive integral values. Let $f(x) = x^2 + 3$. Examination of $\langle f(m) \rangle =$ 3, 4, 7, 12, 17, 28, 39, 52, 67, 84, ... leads one to conjecture that

$$f(x)f(x+1) = f[x(x+1) + 3] - f(x^2 + x + 3).$$

This is easily proved. Using this property and the above relation between m and n, one has

$$f(n) = f(m^2 + m + 2)f(m^2 + m + 3) = f(m^2 + m + 2)f(m)f(m+1).$$

Thus $p|f(n)$ with p prime implies that $p|f(k)$ with k equal to m, $m + 1$, or $m^2 + m + 2$. Since each of these possibilities for k satisfies $k^2 < n$, the desired result follows.

B-4.

Let $M = M(a, b, c)$ denote the 5 by 7 matrix (a_{ij}) with

$$a_{11} = a,\ a_{22} = a_{33} = a_{44} = a_{55} = b,\ a_{16} = a_{27} = c,$$

and $a_{ij} = 0$ in all other cases. Then the set V of all such M (with a, b, c arbitrary real numbers) is closed under linear combinations. Also, $M(0,0,0)$, $M(1,0,0)$, $M(0,0,1)$, $M(0,1,0)$, and $M(1,1,0)$ have ranks 0, 1, 2, 4, and 5, respectively. But no M in V has rank 3 since $b \neq 0$ implies that the rank is 4 or 5 and $b = 0$ forces the rank to be 0, 1, or 2.

B-5.

If n has d digits in base 2, $2^{d-1} \leq n$ and so

$$B(n) \leq d \leq 1 + \log_2 n.$$

This readily implies that $\Sigma_{n=1}^{\infty}[B(n)/n(n+1)]$ converges to a real number S. Hence the manipulations below with convergent series are allowable in the two solutions which follow.

Each n is uniquely expressible as $n_0 + 2n_1 + 2^2 n_2 + \cdots$ with each n_i in $\langle 0, 1\rangle$ (and with $n_i = 0$ for all but a finite set of i). Since

$$1 + 2 + 2^2 + \cdots + 2^{i-1} = 2^i - 1,$$

one sees that $n_i = 1$ if and only if n is of the form $k + 2^i + 2^{i+1}j$ with k in $\langle 0, 1, \ldots, 2^i - 1\rangle$ and j in $\langle 0, 1, 2, \ldots \rangle$. Thus

$$S = \sum_{n=1}^{\infty} \frac{1}{n(n+1)} \sum_{i=0}^{\infty} n_i$$

$$= \sum_{i=0}^{\infty} \sum_{j=0}^{\infty} \sum_{k=0}^{2^i - 1} \frac{1}{(k + 2^i + 2^{i+1}j)(1 + k + 2^i + 2^{i+1}j)}.$$

Using $1/s(s+1) = 1/s - 1/(s+1)$, the innermost sum telescopes and

$$S = \sum_{i=0}^{\infty} \sum_{j=0}^{\infty} \left[\frac{1}{2^i(1+2j)} - \frac{1}{2^i(2+2j)} \right] = \sum_{i=0}^{\infty} \frac{1}{2^i} \sum_{j=0}^{\infty} (-1)^j \frac{1}{j}.$$

Since it is well known that $1 - \frac{1}{2} + \frac{1}{3} - \frac{1}{4} + \cdots = \ln 2$,

$$S = \left(\sum_{i=0}^{\infty} 2^{-i} \right) \ln 2 = 2 \ln 2 = \ln 4$$

and e^S is the rational number 4.

Alternatively, we note that $B(2m) = B(m)$, $B(2m+1) = 1 + B(2m) = 1 + B(m)$.

Then

$$S = \sum_{n=1}^{\infty} \frac{B(n)}{n(n+1)} = \sum_{m=0}^{\infty} \frac{B(2m+1)}{(2m+1)(2m+2)} + \sum_{m=1}^{\infty} \frac{B(2m)}{2m(2m+1)}$$

$$= \sum_{m=0}^{\infty} \frac{1 + B(m)}{(2m+1)(2m+2)} + \sum_{m=1}^{\infty} \frac{B(m)}{2m(2m+1)}$$

$$= \sum_{m=0}^{\infty} \frac{1}{(2m+1)(2m+2)} + \sum_{m=1}^{\infty} B(m) \left[\frac{1}{2m(2m+1)} + \frac{1}{(2m+1)(2m+2)} \right]$$

$$= \ln 2 + \frac{1}{2} \sum_{m=1}^{\infty} \frac{B(m)}{m(m+1)} = \ln 2 + \frac{S}{2}.$$

Hence $S/2 = \ln 2$, $S = \ln 4$, and $\exp(S)$ is the rational number 4.

B-6.

Let $L = L(P)$ be the perimeter of P. One sees that $H(P)$ consists of the region bounded by P, the regions bounded by rectangles whose bases are the sides of P and whose altitudes equal 1, and sectors of unit circles which can be put together to form one unit circle. Hence

$$F(P) = (L/2) + L + \pi = \pi + 3L/2.$$

If A and B are two consecutive vertices of P, the contribution of side AB to the double integral

I is double the area of the region (of the figure) bounded by the unit semicircles with centers at A and B and segments CD and EF such that $ABCD$ and $ABEF$ are rectangles and $|AD| = 1 = |AF|$.

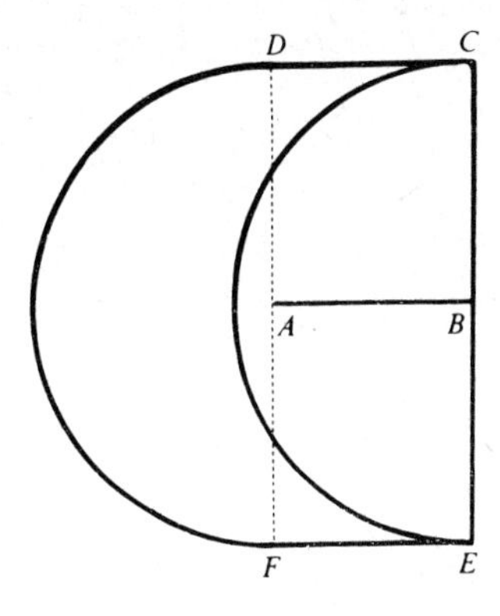

FIG. 1.

One doubles this area because there is a symmetric region bounded by CD, EF, and the other halves of the unit circles centered at A and B. The overlap of the two regions counts twice. By Cavalieri's slicing principle, this contribution of side AB to I is 4 times the length of AB. Hence $I = 4L$ and

$$\frac{I}{F(P)} = \frac{4L}{\pi + 3L/2} = \frac{8}{3 + (2\pi/L)}.$$

One can make L arbitrarily large (e.g., by letting P be a triangle with two angles arbitrarily close to right angles). Hence the desired least upper bound is $8/3$.

THE FORTY-THIRD WILLIAM LOWELL PUTNAM MATHEMATICAL COMPETITION

December 4, 1982

A-1.

Let T consist of the points inside or on the triangle with vertices at $(0,3), (-1,4), (1,4)$ and let U be the set of points inside or on the triangle with vertices at $(0,0), (-4,4), (4,4)$. Then T and V overlap only on boundary points and their union is U. The centroids of T and U are $(0,11/3)$ and $(0,8/3)$, respectively. The areas of T, V, and U are 1, 15, and 16, respectively. Using weighted averages with the areas as weights, one has

$$1 \cdot 0 + 15\bar{x} = 16 \cdot 0, \qquad 1 \cdot \tfrac{11}{3} + 15\bar{y} = 16 \cdot \tfrac{8}{3}.$$

It follows that $\bar{x} = 0$, $\bar{y} = 13/5$.

A-2.

Since $x = \log_n 2 > 0$, $B_n(x) = 1^x + 2^x + \cdots + n^x \leqslant n \cdot n^x$ and

$$0 \leqslant \frac{B_n(\log_n 2)}{(n\log_2 n)^2} \leqslant \frac{n \cdot n^{\log_n 2}}{(n\log_2 n)^2} = \frac{2}{n(\log_2 n)^2}.$$

As $\sum_{n=2}^{\infty}[2/n(\log_2 n)^2]$ converges by the integral test, the given series converges by the comparison test.

A-3.

$$\int_0^\infty \frac{\operatorname{Arctan}(\pi x) - \operatorname{Arctan} x}{x}\,dx = \int_0^\infty \frac{1}{x}\operatorname{Arctan}(ux)\Big|_{u=1}^{u=\pi} dx$$

$$= \int_0^\infty \int_1^\pi \frac{1}{1+(xu)^2}\,du\,dx = \int_1^\pi \int_0^\infty \frac{1}{1+(xu)^2}\,dx\,du$$

$$= \int_1^\pi \frac{1}{u}\cdot\frac{\pi}{2}\,du = \frac{\pi}{2}\ln\pi.$$

A-3. *Alternate solution*

Let

$$f(a) = \int_0^\infty \frac{\operatorname{Arctan}(ax) - \operatorname{Arctan} x}{x}\,dx.$$

Then

$$f'(a) = \int_0^\infty \frac{1}{x}\left(\frac{x}{1+a^2x^2}\right)dx = \frac{1}{a}\int_0^\infty \frac{dy}{1+y^2} = \frac{1}{a}\cdot\frac{\pi}{2}.$$

Therefore $f(a) = (\pi/2)\ln a + C$. But $C = 0$ since $f(1) = 0$. Hence $f(\pi) = (\pi/2)\ln \pi$.

A-4.

The differential equations imply that

$$y^3y' + z^3z' = z'y' - y'z' = 0$$

and hence that $y^4 + z^4$ is constant. This and the initial conditions give $y^4 + z^4 = 1$. Thinking of x as a time variable and (y, z) as the coordinates of a point in a plane, this point moves on the curve $y^4 + z^4 = 1$ with speed

$$\left[(y')^2 + (z')^2\right]^{1/2} = \left(z^6 + y^6\right)^{1/2}.$$

At any time, either y^4 or z^4 is at least $\frac{1}{2}$ and so the speed is at least $\{[(\frac{1}{2})^{1/4}]^6\}^{1/2}$. Hence the point will go completely around the finite curve in some time L. As the speed depends only on y and z (and not on x), the motion is periodic with period L.

A-5.

We are given that

$$r = 1 - \frac{a}{b} - \frac{c}{d} = \frac{bd - ad - bc}{bd} > 0.$$

Thus $bd - ad - bc$ is a positive integer and so $r \geqslant 1/bd$. We may assume without loss of generality that $b \leqslant d$. If $b \leqslant d \leqslant 1983$, $r \geqslant 1983^{-2} > 1983^{-3}$. Since $a + c \leqslant 1982$, if $1983 \leqslant b \leqslant d$, one has

$$r \geqslant 1 - \frac{a}{1983} - \frac{c}{1983} \geqslant 1 - \frac{1982}{1983} = \frac{1}{1983} > \frac{1}{1983^3}.$$

The remaining case is that with $b < 1983 < d$. Then the d that minimizes r for fixed a, b, c is $1 + [bc/(b - a)]$, where $[x]$ is the greatest integer in x. This d is at most $1983b$ since $b - a \geqslant 1$ and $c < 1982$ and thus

$$r \geqslant \frac{1}{bd} \geqslant \frac{1}{1983b^2} > \frac{1}{1983^3}.$$

Hence we have the desired inequality in all cases.

A-6.

We disprove the assertion. Let $y_n = 1/(n + 1)\ln(n + 1)$. Then $\Sigma_{n=1}^{\infty}(-1)^{n+1}y_n$ converges to some $g > 0$ since $y_n \to 0$ as $n \to \infty$ and $y_1 > y_2 > \cdots$. Let $x_n = (-1)^{n+1}y_n/g$. Then conditions (i) and (iii) are satisfied. Let $a_0, a_1, \ldots$ be positive integers to be made more definite later. Let $b_0 = 0$ and $b_{i+1} = b_i + 4a_i$ for $i = 0, 1, \ldots$. The bijection σ is defined as follows:

$$\sigma(n) = 2n - 1 - b_i \quad \text{for} \quad b_i < n \leqslant b_i + 2a_i,$$

$$\sigma(n) = 2n - b_{i+1} \quad \text{for} \quad b_i + 2a_i < n \leqslant b_{i+1}.$$

Then $0 < \sigma(n) < 2n$ and hence $|\sigma(n) - n| < n$. Thus

$$|\sigma(n) - n| \cdot |x_n| < \frac{n}{g(n + 1)\ln(n + 1)}$$

which implies condition (ii). Let $C(n) = \Sigma_{i=1}^{n}x_i$ and $D(n) = \Sigma_{i=1}^{n}x_{\sigma(i)}$. Then

$$\text{(A)} \qquad D(b_i + 2a_i) - C(b_i + 2a_i) = \frac{1}{g}\sum_{j=1}^{a_i} y_{b_i+2j} + \frac{1}{g}\sum_{k=1}^{a_i} y_{b_i+2a_i+2k-1}.$$

Since $y_2 + y_4 + y_6 + \cdots$ diverges to $+\infty$ by the integral test, the a_i can be chosen large enough so that the first sum in (A) exceeds 1 for each i. Then $D(n) > 1 + C(n)$ for an unbounded sequence of n's. Hence $D(n)$ and $C(n)$ cannot converge to the same limit.

B-1.

The smallest n is 2. Let D be the midpoint of side AB. Cut ΔAMB along DM. Then ΔBMD can be placed alongside ΔADM, with side BD atop side AD, so as to form a triangle congruent to ΔAMC. Since ΔAMB need not be congruent to ΔAMC in a general ΔABC, there is no method with $n = 1$.

B-2.

Let $r = \sqrt{x^2 + y^2}$, $R(m, n) = \{(x, y) : m^2 + n^2 \leqslant x^2 + y^2\}$, and

$$I = \int_{-\infty}^{\infty}\int_{-\infty}^{\infty} A(x, y)e^{-x^2-y^2}\,dx\,dy.$$

Let Σ and Σ' denote sums over all integers m and over all integers n, respectively. Then

$$\begin{aligned}
I &= \Sigma\Sigma'\iint_{R(m,n)} e^{-x^2-y^2}\,dx\,dy\\
&= \Sigma\Sigma'\int_0^{2\pi}\int_{\sqrt{m^2+n^2}}^{\infty} e^{-r^2} r\,dr\,d\theta\\
&= \Sigma\Sigma'\int_0^{2\pi}\left[-\tfrac{1}{2}e^{-r^2}\right]_{\sqrt{m^2+n^2}}^{\infty} d\theta\\
&= \Sigma\Sigma'\int_0^{2\pi}\tfrac{1}{2}e^{-m^2-n^2}\,d\theta\\
&= \Sigma\Sigma'\pi e^{-m^2-n^2} = \pi\left(\Sigma e^{-m^2}\right)\left(\Sigma' e^{-n^2}\right) = \pi(2g-1)^2.
\end{aligned}$$

B-3.

Let $a(n) = [\sqrt{n+1}\,]$ and $b(n) = [\sqrt{2n}\,]$, where $[x]$ is the greatest integer in x. For t in $\{1, 2, \ldots, a(n)\}$, there are $t^2 - 1$ ordered pairs (c, d) with c and d in $X = \{1, 2, \ldots, n\}$ and $c + d = t^2$. For t in $\{1 + a(n), 2 + a(n), \ldots, b(n)\}$, there are $2n + 1 - t^2$ ordered pairs (c, d) with c and d in X and $c + d = t^2$. Hence the total number $F(n)$ of favorable (c, d) is

$$\begin{aligned}
F(n) &= \sum_{t=1}^{a(n)} (t^2 - 1) + \sum_{t=1+a(n)}^{b(n)} (2n + 1 - t^2)\\
&= \left(2\sum_{t=1}^{a(n)} t^2\right) - \left(\sum_{t=1}^{b(n)} t^2\right) - a(n) + [b(n) - a(n)](2n+1)\\
&= \frac{2a(n)[1 + a(n)][1 + 2a(n)]}{6} - \frac{b(n)[1 + b(n)][1 + 2b(n)]}{6}\\
&\quad - 2(n+1)a(n) + (2n+1)b(n).
\end{aligned}$$

Since $p_n = F(n)/n^2$,

$$\begin{aligned}
\lim_{n\to\infty}\left(p_n\sqrt{n}\right) &= \lim_{n\to\infty} F(n)/n^{3/2}\\
&= \frac{2\cdot 2}{6}\lim_{n\to\infty}\left(\frac{a(n)}{\sqrt{n}}\right)^3 - \frac{2}{6}\lim_{n\to\infty}\left(\frac{b(n)}{\sqrt{n}}\right)^3 - 2\lim_{n\to\infty}\frac{a(n)}{\sqrt{n}} + 2\lim_{n\to\infty}\frac{b(n)}{\sqrt{n}}\\
&= \frac{2}{3} - \frac{1}{3}(\sqrt{2})^3 - 2 + 2\sqrt{2} = \frac{4}{3}(\sqrt{2} - 1).
\end{aligned}$$

B-4.

Let $P_k = (n_1 + k)(n_2 + k) \cdots (n_s + k)$. We are given that $P_0 | P_k$ for all integers k.

(a) $P_0 | P_{-1}$ and $P_0 | P_1$ together imply $P_0^2 | (P_{-1}P_1)$ or $(n_1^2 n_2^2 \cdots n_s^2) | [(n_1^2 - 1)(n_2^2 - 1) \cdots (n_s^2 - 1)]$.

No n_i can be zero since $P_k \neq 0$ for k sufficiently large. Thus, for each i, $n_i^2 \geq 1$ and $n_i^2 > n_i^2 - 1 \geq 0$. Hence $P_0^2 > P_{-1}P_1 \geq 0$. This and $P_0^2 | (P_{-1}P_1)$ imply $P_{-1}P_1 = 0$. Then for some i, $|n_i| = 1$.

(b) P_k is a polynomial in k of degree s. Since P_0 divides each P_i, P_0 also divides the nth difference

$$\sum_{i=0}^{s} (-1)^i \binom{s}{i} P_i = s!.$$

Since $P_0 > 0$, this means that $P_0 \leq s!$. As P_0 is a product of s distinct positive integers, it follows that

$$\{n_1, n_2, \ldots, n_s\} = \{1, 2, \ldots, s\}.$$

B-5.

Since the derivative of $x^{1/x}$ is negative for $x > e$,

(1) $$a^b > b^a \text{ when } e \leq a < b.$$

The u's are defined so that $u_0 = e$ and

(2) $$x = (u_0)^{u_1} = (u_1)^{u_2} = (u_2)^{u_3} = \cdots.$$

Hence

(3) $$u_{n+1} = (u_n \ln u_{n-1})/\ln u_n.$$

As $x > e^e$, $u_1 = \ln x > e = u_0$. Now $u_1 > u_0$ implies $\ln u_1 > \ln u_0$ and then (3) with $n = 1$ implies $u_2 < u_1$. Also, (2) and (1) imply $(u_1)^{u_2} = (u_0)^{u_1} > (u_1)^{u_0}$, which gives us $u_2 > u_0$. Now $u_2 < u_1$ and (3) with $n = 2$ imply $u_3 > u_2$. Also (2) and (1) imply $(u_2)^{u_3} = (u_1)^{u_2} < (u_2)^{u_1}$ and hence $u_3 < u_1$. Similarly, $u_2 < u_4 < u_3$. Then an easy induction shows that

$$e < u_{2n} < u_{2n+2} < u_{2n+1} < u_{2n-1} \text{ for } n = 1, 2, \ldots.$$

Thus the monotonic bounded sequences $u_0, u_2, u_4, \ldots$ and $u_1, u_3, u_5, \ldots$ have limits a and b, respectively, with $e < a \leq b$. Also

$$a^b = \lim_{n \to \infty} (u_{2n})^{u_{2n+1}} = x = \lim_{n \to \infty} (u_{2n-1})^{u_{2n}} = b^a.$$

Then $a^b = b^a$, $e \leq a \leq b$, and (1) imply $a = b$. Hence $\lim_{n \to \infty} u_n$ exists and is the unique real number $g = g(x)$ with $g > e$ and $g^g = x$. Since $f(y) = y^y$ is continuous and strictly increasing for $y \geq e$, its inverse function $g(x)$ is also continuous.

B-6.

Let $s = (a + b + c)/2$, $t = s - a$, $u = s - b$, $v = s - c$ and similarly for the primed letters. Using Heron's Formula, the inequality to be proved will follow from

(A) $$\sqrt[4]{stuv} + \sqrt[4]{s't'u'v'} \leq \sqrt[4]{(s + s')(t + t')(u + u')(v + v')}$$

for positive $s, t, u, v, s', t', u', v'$. A simpler analogous inequality that might be helpful is

(B) $$\sqrt{xy} + \sqrt{x'y'} \leq \sqrt{(x + x')(y + y')} \quad \text{for} \quad x, y, x', y' \text{ positive.}$$

First we note that (B) follows from the Cauchy Inequality applied to the vectors $(\sqrt{x}, \sqrt{x'})$ and $(\sqrt{y}, \sqrt{y'})$ [and also follows from $(\sqrt{xy'} - \sqrt{x'y})^2 \geq 0$ or from the Inequality on the Means applied to

xy' and $x'y$]. Using (B) with $x = \sqrt{st}$, $x' = \sqrt{s't'}$, $y = \sqrt{uv}$, $y' = \sqrt{u'v'}$ and reapplying (B) to the new right side, one has

$$\sqrt[4]{stuv} + \sqrt[4]{s't'u'v'} \leqslant \sqrt{(\sqrt{st} + \sqrt{s't'})(\sqrt{uv} + \sqrt{u'v'})}$$

$$\leqslant \sqrt{\sqrt{(s+s')(t+t')}\sqrt{(u+u')(v+v')}}\,.$$

Since here the rightmost part equals the right side of (A), we have proved (A).

Equality holds in (B) if and only if $\sqrt{x} : \sqrt{x'} = \sqrt{y} : \sqrt{y'}$ and this holds if and only if $x : x' = y : y'$. Hence equality occurs in (A) if and only if $s : t : u : v = s' : t' : u' : v'$. It follows that equality occurs in the original inequality if and only if a, b, c are proportional to a', b', c'.

THE FORTY-FOURTH WILLIAM LOWELL PUTNAM MATHEMATICAL COMPETITION

December 3, 1983

A-1.

For d and m in $Z^+ = \{1,2,3,\dots\}$, let $d|m$ denote that d is an integral divisor of m. For m in Z^+, let $\tau(m)$ be the number of d in Z^+ such that $d|m$. The number of n in Z^+ such that $n|a$ or $n|b$ is

$$\tau(a) + \tau(b) - \tau(\gcd(a,b)).$$

Also $\tau(p^s q^t) = (s+1)(t+1)$ for p, q, s, t in Z^+ with p and q distinct primes. Thus the desired count is

$$\begin{aligned}\tau(2^{40}\cdot 5^{40}) + \tau(2^{60}\cdot 5^{30}) - \tau(2^{40}\cdot 5^{30}) &= 41^2 + 61\cdot 31 - 41\cdot 31\\ &= 1681 + 620 = 2301.\end{aligned}$$

A-2.

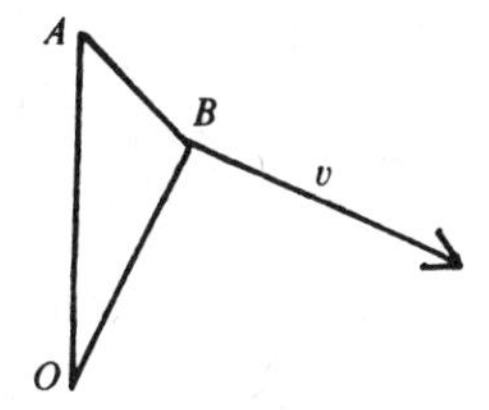

Let OA be the long hand and OB be the short hand. We can think of OA as fixed and OB as rotating at constant speed. Let v be the vector giving the velocity of point B under this assumption. The rate of change of the distance between A and B is the component of v in the direction of AB. Since v is orthogonal to OB and the magnitude of v is constant, this component is maximal when $\measuredangle OBA$ is a right angle, i.e., when the distance AB is $\sqrt{4^2 - 3^2} = \sqrt{7}$.

Alternatively, let x be the distance AB and $\theta = \measuredangle AOB$. By the Law of Cosines,

$$x^2 = 3^2 + 4^2 - 2\cdot 3\cdot 4\cos\theta = 25 - 24\cos\theta.$$

Since $d\theta/dt$ is constant, we may assume units chosen so that θ is also time t. Now

$$2x\frac{dx}{d\theta} = 24\sin\theta, \quad \frac{dx}{d\theta} = \frac{12\sin\theta}{\sqrt{25 - 24\cos\theta}}.$$

Since $dx/d\theta$ is an odd function of θ, $|dx/ds|$ is a maximum when $dx/d\theta$ is a maximum or a minimum. Since dx/ds is a periodic differentiable function of θ, $d^2x/ds^2 = 0$ at the extremes for dx/ds. For such θ,

$$12\cos\theta = x\frac{d^2x}{d\theta^2} + \left(\frac{dx}{d\theta}\right)^2 = \left(\frac{dx}{d\theta}\right)^2 = \frac{144\sin^2\theta}{x^2}.$$

Then

$$x^2 = \frac{12\sin^2\theta}{\cos\theta} = \frac{12 - 12\cos^2\theta}{\cos\theta} = 25 - 24\cos\theta,$$

and it follows that

$$12\cos^2\theta - 25\cos\theta + 12 = 0.$$

The only allowable solution for $\cos\theta$ is $\cos\theta = 3/4$ and hence $x = \sqrt{25 - 24\cos\theta} = \sqrt{25 - 18} = \sqrt{7}$.

A-3.

$$F(n) = 1 + 2n + 3n^2 + \cdots + (p-1)n^{p-2},$$
$$nF(n) = n + 2n^2 + \cdots + (p-2)n^{p-2} + (p-1)n^{p-1}.$$

Hence $(1-n)F(n) = (1 + n + n^2 + \cdots + n^{p-2}) - (p-1)n^{p-1}$ and similarly

$$(1-n)^2F(n) = 1 - n^{p-1} - (1-n)(p-1)n^{p-1} = 1 - p\cdot n^{p-1} + (p-1)n^p.$$

Modulo p, $n^p \equiv n$ by the Little Fermat Theorem and so $(1-n)^2F(n) \equiv 1 - n$. If neither a nor b is congruent to 1 (mod p), $1 - a \not\equiv 1 - b$ and there are distinct reciprocals $(1-a)^{-1}$ and $(1-b)^{-1}$(mod p); then

$$f(a) \equiv (1-a)^{-1}, f(b) \equiv (1-b)^{-1}, f(a) \not\equiv f(b) (\text{mod } p).$$

If one of a and b, say a, is congruent to 1, then $b \not\equiv 0$ (mod p) and so $f(b) \equiv (1-b)^{-1} \not\equiv 0$ (mod p) while

$$f(a) = 1 + 2 + \cdots + (p-1) = p(p-1)/2 \equiv 0 (\text{mod } p).$$

A-4.

Let $\binom{m}{r} = 0$ for $r > m$ and for $r < 0$. For $i = 0, 1, 2$ let

$$T_i(m) = \binom{m}{i} - \binom{m}{i+3} + \binom{m}{i+6} - \binom{m}{i+9} + \cdots.$$

We note that $S(m) = T_2(m) + 1$. Since $\binom{n}{r} = \binom{n-1}{r} + \binom{n-1}{r-1}$,

$$T_2(m) = T_2(m-1) + T_1(m-1),\ T_1(m) = T_1(m-1) + T_0(m-1),$$
$$T_0(m) = T_0(m-1) - T_2(m-1).$$

Let the backwards difference operator ∇ be given by $\nabla f(n) = f(n) - f(n-1)$. Then

$$\nabla T_2(m) = T_1(m-1),\ \nabla T_1(m) = T_0(m-1),\ \nabla T_0(m) = -T_2(m-1).$$

These imply that

$$\nabla^3 T_2(m) = \nabla^2 T_1(m-1) = \nabla T_0(m-2) = -T_2(m-3) \text{ for } m \geqslant 3.$$

Expanding $\nabla^3 T_2(m)$, this gives us

(R) $$T_2(m) = 3[T_2(m-1) - T_2(m-2)] \text{ for } m \geqslant 3.$$

When $m = 6k - 1$ with $k \geqslant 1$, we have $m \geqslant 5$. It then follows from (R) that $T_2(m) \equiv 0$ (mod 3) and hence $S(m) \equiv 1$ (mod 3). Thus $S(m) \neq 0$.

A-5.

Inductively we define a sequence of integers $3 = a_1, a_2, a_3, \ldots$ and associated intervals $I_n = [(a_n)^{1/n}, (1 + a_n)^{1/n})$ such that $a_n \geqslant 3^n$, $a_n \equiv n$ (mod 2), the sequence $\{(a_n)^{1/n}\}$ is nondecreasing, and $I_n \supseteq I_{n+1}$. When this has been done, $\{(a_n)^{1/n}\}$, being nondecreasing and bounded, will have a limit u which is in I_n for all n. Then $(a_n)^{1/n} \leqslant u < (1 + a_n)^{1/n}$ will imply that $a_n \leqslant u^n < 1 + a_n$ and so $[u^n] = a_n \equiv n$(mod 2) for all n.

Let $a_1 = 3$. Then $I_1 = [3,4)$. Let us assume that we have $a_1, a_2, \ldots, a_k$ and $I_1, I_2, \ldots, I_k$ with the desired properties. Let

$$J_k = \left[(a_k)^{(k+1)/k}, (1 + a_k)^{(k+1)/k}\right).$$

Then x is in I_k if and only if x^{k+1} is in J_k. The length of J_k is

$$(1 + a_k)^{(k+1)/k} - (a_k)^{(k+1)/k} \geqslant (1 + a_k - a_k)(a_k)^{1/k} = a_k^{1/k} \geqslant (3^k)^{1/k} = 3.$$

Since the length of J_k is at least 3, J_k contains an interval $L_k = [a_{k+1}, 1 + a_{k+1})$ for some integer a_{k+1} which is congruent to $k+1$ (mod 2). Let

$$I_{k+1} = \left[(a_{k+1})^{1/(k+1)}, (1 + a_{k+1})^{1/(k+1)}\right).$$

Since $x \in I_k$ if and only if $x^{k+1} \in J_k$, $x \in I_{k+1}$ if and only if $x^{k+1} \in L_k$, and $J_k \supseteq L_k$, one sees that $I_k \supseteq I_{k+1}$. Also

$$a_{k+1} \geqslant (a_k)^{(k+1)/k} = \left[(a_k)^{1/k}\right]^{k+1} \geqslant 3^{k+1}.$$

This completes the inductive step and shows that the desired u exists.

A-6.

Under the change of variables $s = u - v$ and $t = u + v$, with the Jacobian $\partial(u,v)/\partial(s,t) = 1/2$, $F(x)$ becomes $I(x)/E(x)$ where

$$\begin{aligned} I(x) &= \int_0^x \int_{-t}^t \exp\left\{\left(\frac{t+s}{2}\right)^3 + \left(\frac{t-s}{2}\right)^3\right\} ds\, dt \\ &= \int_0^x \int_{-t}^t \exp\left(\frac{1}{4}t^3 + \frac{3}{4}ts^2\right) ds\, dt \end{aligned}$$

and $E(x) = 2x^{-4}\exp(x^3)$. Since $I(x)$ and $E(x)$ go to $+\infty$ as x goes to $+\infty$, one can use L'Hôpital's Rule and we have $\lim_{x\to\infty} F(x) = \lim_{x\to\infty}(I'/E')$ where

$$I' = \int_{-x}^x \exp\left(\frac{1}{4}x^3 + \frac{3}{4}xs^2\right) ds = \exp(x^3/4)\int_{-x}^x \exp(3xs^2/4)\, ds$$

and $E' = (6x^{-2} - 8x^{-5})\exp(x^3)$. In the integral for I', make the change of variable $s = w/\sqrt{x}$, $ds = dw/\sqrt{x}$, to obtain

$$I' = \frac{\exp(x^3/4)}{\sqrt{x}} \int_{-x\sqrt{x}}^{x\sqrt{x}} \exp(3w^2/4)\, dw.$$

Now

$$\lim_{x\to\infty} F(x) = \lim_{x\to\infty} \frac{I'}{E'} = \lim_{x\to\infty} \frac{\int_{-x\sqrt{x}}^{x\sqrt{x}} \exp(3w^2/4)\, dw}{(6x^{-3/2} - 8x^{-9/2})\exp(3x^3/4)}.$$

We can, and do, use L'Hôpital's rule again to obtain

$$\lim_{x\to\infty} F(x) = \lim_{x\to\infty} \frac{2(3/2)x^{1/2}\exp(3x^2/4)}{[(27/2)x^{1/2} + \cdots]\exp(3x^2/4)} = \frac{2}{9}.$$

B-1.

The diameter of S must be 4 and S must be centered at the center of C. The set of points inside C nearer to v than to another vertex w is the part of that half-space, bounded by the perpendicular bisector of the segment vw, containing v which lies within C. The intersection of these sets is a cube C' bounded by the three facial planes of C through v and the three planes which are perpendicular bisectors of the edges of C at v. These last 3 planes are planes of symmetry for C and S. Hence R is one of 8 disjoint congruent regions whose union is the set of points between S and C, excepting those on the 3 planes of symmetry. Therefore

$$8\,\mathrm{vol}(R) = \mathrm{vol}(C) - \mathrm{vol}(S) = 4^3 - \frac{4\pi}{3}\cdot 2^3,$$

$$\mathrm{vol}(R) = 8 - \frac{4\pi}{3}.$$

B-2.

A representation for $2n$ is of the form

$$2n = e_0 + 2e_1 + 4e_2 + \cdots + 2^k e_k,$$

the e_i in $\{0,1,2,3\}$, and with e_0 in $\{0,2\}$. Then $e_1 + 2e_2 + \cdots + 2^{k-1}e_k$ is a representation for n if $e_0 = 0$ and is a representation for $n-1$ if $e_0 = 2$. Since all representations for n and $n-1$ can be obtained this way,

$$C(2n) = C(n) + C(n-1).$$

Similarly, one finds that

$$C(2n+1) = C(n) + C(n-1) = C(2n).$$

Since $C(1) = 1$ and $C(2) = 2$, an easy induction now shows that $C(n) = [1 + n/2]$.

B-3.

To satisfy the equation, each y_i must have at least 3 derivatives. Here Σ will be a sum with i running over 1, 2, 3. We have $\Sigma y_i^2 = 1$ and $\Sigma(y_i')^2 = f$. Differentiating, one has $\Sigma 2y_iy_i' = 0$ and $\Sigma 2y_i'y_i'' = f'$. Differentiating $\Sigma y_iy_i' = 0$ leads to $\Sigma y_iy_i'' + \Sigma(y_i')^2 = 0$ so $\Sigma y_iy_i'' = -f$. Differentiating this gives us $\Sigma y_i'y_i'' + \Sigma y_iy_i''' = -f'$. This and $\Sigma y_i'y_i'' = f'/2$ leads to $\Sigma y_iy_i''' = -3f'/2$. Multiplying each term of

$$y_i''' + py_i'' + qy_i' + ry_i = 0$$

by y_i and summing gives us

$$-3f'/2 - pf + q\cdot 0 + r = 0.$$

Thus $f' + (2/3)pf = (2/3)r$ and so $A = 2/3 = B$.

B-4.

We can let $m = k^2 + j$, where k and j are integers with $0 \leqslant j \leqslant 2k$, since the next square after k^2 is $k^2 + 2k + 1$; let this j be the *excess* for m. We note that $[\sqrt{m}] = k$ and $f(m) = k^2 + j + k$. If the excess j is 0, m is already a square. Let A consist of the m's with excess j satisfying $0 \leqslant j \leqslant k$ and B consist of the m's with $k < j \leqslant 2k$. If m is in B,

$$f(m) = k^2 + j + k = (k+1)^2 + (j - k - 1),$$

with the excess $j - k - 1$ for $f(m)$ satisfying $0 \leqslant j - k - 1 \leqslant k + 1$, and hence $f(m)$ is either a square or is in A. Thus it suffices to deal with the case in which m is in A. Then $[\sqrt{m+k}] = k$ and

$$f^2(m) = f(f(m)) = f(m+k) = m + 2k = (k+1)^2 + (j-1).$$

Hence $f^2(m)$ is either a square or an integer in A with excess smaller than that of m. Continuing, one sees that $f^r(m)$ is a square for some r with $0 \leqslant r \leqslant 2j$.

B-5.

By definition of a_n and $\|u\|$,

$$a_n = \sum_{k=1}^{n-1} \frac{1}{n}\left[\int_{2n/(2k+1)}^{n/k}\left(\frac{n}{x} - k\right)dx + \int_{n/(k+1)}^{2n/(2k+1)}\left(k+1-\frac{n}{x}\right)dx\right]$$

$$= \sum_{k=1}^{n-1}\left[\ln\frac{2k+1}{2k} - \frac{1}{2k+1} + \frac{1}{2k+1} - \ln\frac{2k+2}{2k+1}\right]$$

$$= \ln\prod_{k=1}^{n-1}\frac{(2k+1)^2}{2k(2k+2)} = \ln\left[\frac{3}{2}\cdot\frac{3}{4}\cdot\frac{5}{4}\cdot\frac{5}{6}\cdots\frac{(2n-1)}{(2n-2)}\cdot\frac{(2n-1)}{2n}\right].$$

Since

$$\frac{2}{1}\cdot\frac{2}{3}\cdot\frac{4}{3}\cdot\frac{4}{5}\cdot\frac{6}{5}\cdot\frac{6}{7}\cdots = \frac{\pi}{2}$$

and $\ln x$ is continuous for $x > 0$, $\lim_{n\to\infty} a_n = \ln(4/\pi)$.

B-6.

Since $r \neq 1$ and $r^m - 1 = (r-1)(r^{m-1} + r^{m-2} + \cdots + 1) = 0$, one has $r^{m-1} + r^{m-2} + \cdots + 1 = 0$ and so

$$-1 = r(1 + r + r^2 + \cdots + r^{m-2}),$$
$$-1 = r(1+r)(1+r^2)(1+r^4)\cdots(1+r^{(m-1)/2}),$$
$$-1 = (r+r^2)(1+r^2)(1+r^4)\cdots(1+r^{(m-1)/2}).$$

Since $r + r^2 = r^{m+1} + r^2$ with $m + 1 = 2(2^{k-1}+1)$, each of the factors in the last expression for -1 is a sum of two squares. Their product can be expressed as a sum of two squares by repeated application of the identity

$$(a^2+b^2)(c^2+d^2) = (ac-bd)^2 + (ad+bc)^2.$$

This converts -1 into $P^2 + Q^2$ with each of P and Q a polynomial in r with integer coefficients.

THE FORTY-FIFTH WILLIAM LOWELL PUTNAM MATHEMATICAL COMPETITION

December 1, 1984

A-1.

The set B can be partitioned into the following sets:

(i) A itself, of volume abc;
(ii) two $a \times b \times 1$ bricks, two $a \times c \times 1$ bricks, and two $b \times c \times 1$ bricks, of total volume $2ab + 2ac + 2bc$;
(iii) four quarter-cylinders of length a and radius 1, four quarter-cylinders of length b and radius 1, and four quarter-cylinders of length c and radius 1, of total volume $(a + b + c)\pi$;
(iv) eight spherical sectors, each consisting of one-eighth of a sphere of radius 1, of total volume $4\pi/3$.

Hence the volume of B is

$$abc + 2(ab + ac + bc) + \pi(a + b + c) + \frac{4\pi}{3}.$$

A-2.

Let $S(n)$ denote the nth partial sum of the given series. Then

$$S(n) = \sum_{k=1}^{n}\left[\frac{3^k}{3^k - 2^k} - \frac{3^{k+1}}{3^{k+1} - 2^{k+1}}\right] = 3 - \frac{3^{n+1}}{3^{n+1} - 2^{n+1}},$$

and the series converges to $\lim_{n\to\infty} S(n) = 2$.

A-3.

Let $N = M_n]_{x=a}$. N has rank 2, so that 0 is an eigenvalue of multiplicity $2n - 2$. Let $\mathbf{e}$ denote the $2n \times 1$ column vector of 1's. Notice that $N\mathbf{e} = n(a + b)\mathbf{e}$, and therefore $n(a + b)$ is an eigenvalue. The trace of N is $2na$, and therefore the remaining eigenvalue is $2na - n(a + b) = n(a - b)$. [Note: This corresponds to the eigenvector $\mathbf{f}$, where $f_{i,1} = (-1)^{i+1}$, $i = 1, \dots, 2n$.]
The preceding analysis implies that the characteristic equation of N is

$$\det(N - \lambda I) = \lambda^{2n-2}(\lambda - n(a + b))(\lambda - n(a - b)).$$

Let $\lambda = a - x$. Then

$$\det M_n = \det(N - (a - x)I) = (a - x)^{2n-2}(a - x - n(a + b))(a - x - n(a - b)).$$

It follows that

$$\lim_{x\to a}\frac{\det M_n}{(x - a)^{2n-2}} = \lim_{x\to a}(a - x - n(a + b))(a - x - n(a - b)) = n^2(a^2 - b^2).$$

A-4.

Let $\theta = \text{Arc}\,AB$, $\alpha = \text{Arc}\,DE$, and $\beta = \text{Arc}\,EA$. Then $\text{Arc}\,CD = \pi - \theta$ and $\text{Arc}\,BC = \pi - \alpha - \beta$.

The area of P, in terms of the five triangles from the center of the circle is

$$\frac{1}{2}\sin\theta + \frac{1}{2}\sin(\pi - \theta) + \frac{1}{2}\sin\alpha + \frac{1}{2}\sin\beta + \frac{1}{2}\sin(\pi - \alpha - \beta).$$

This is maximized when $\theta = \pi/2$ and $\alpha = \beta = \pi/3$. Thus, the maximum area is

$$\frac{1}{2}\cdot 1 + \frac{1}{2}\cdot 1 + \frac{1}{2}\frac{\sqrt{3}}{2} + \frac{1}{2}\frac{\sqrt{3}}{2} + \frac{1}{2}\frac{\sqrt{3}}{2} = 1 + \frac{3}{4}\sqrt{3}.$$

A-5.

For $t > 0$, let R_t be the region consisting of all triples (x, y, z) of nonnegative real numbers satisfying $x + y + z \leqslant t$. Let

$$I(t) = \iiint_{R_t} x^1 y^9 z^8 (t - x - y - z)^4 \, dx\, dy\, dz$$

and make the change of variables $x = tu$, $y = tv$, $z = tw$. We see that $I(t) = I(1)t^{25}$.
Let $J = \int_0^\infty I(t)e^{-t}\,dt$. Then

$$J = \int_0^\infty I(1)\,t^{25}e^{-t}\,dt = I(1)\Gamma(26) = I(1)25!.$$

It is also the case that

$$J = \int_{t=0}^\infty \iiint_{R_t} e^{-t}x^1y^9z^8(t - x - y - z)^4 \, dx\, dy\, dz\, dt.$$

Let $s = t - x - y - z$. Then

$$J = \int_0^\infty\int_0^\infty\int_0^\infty\int_0^\infty e^{-s}e^{-x}e^{-y}e^{-z}x^1y^9z^8s^4\,dx\,dy\,dz\,ds = \Gamma(2)\Gamma(10)\Gamma(9)\Gamma(5) = 1!9!8!4!.$$

The integral we desire is $I(1) = J/25! = 1!9!8!4!/25!$.

A-6.

(a) All congruences are modulo 10.

Lemma. $f(5n) \equiv 2^n f(n)$.

Proof. We have

$$(*) \qquad (5n)! = 10^n n! \prod_{i=0}^{n-1} \frac{(5i+1)(5i+2)(5i+3)(5i+4)}{2}.$$

If i is even, then

$$\frac{1}{2}(5i+1)(5i+2)(5i+3)(5i+4) \equiv \frac{1}{2}(1\cdot 2\cdot 3\cdot 4) \equiv 2,$$

and if i is odd, then

$$\frac{1}{2}(5i+1)(5i+2)(5i+3)(5i+4) \equiv \frac{1}{2}(6\cdot 7\cdot 8\cdot 9) \equiv 2.$$

Thus the entire product above is congruent to 2^n. From $(*)$ it is clear that the largest power of 10 dividing $(5n)!$ is the same as the largest power of 10 dividing $10^n n!$, and the proof follows.

We now show by induction on $5^{a_1} + \cdots + 5^{a_k}$ that

$$f(5^{a_1} + \cdots + 5^{a_k}) \equiv 2^{a_1 + \cdots + a_k}$$

(which depends only on $a_1 + \cdots + a_k$ as desired).
This is true for $5^{a_1} + \cdots + 5^{a_k} = 1$, since $f(5^0) \equiv 2^0 \equiv 1$.

CASE 1. All $a_i > 0$. By the lemma and induction,

$$f(5^{a_1} + \cdots + 5^{a_k}) \equiv 2^{5^{a_1-1} + \cdots + 5^{a_k-1}} f(5^{a_1-1} + \cdots + 5^{a_k-1})$$

$$\equiv 2^k \cdot 2^{(a_1-1)+\cdots+(a_k-1)} \qquad (\text{since } 2^{5^i} \equiv 2 \text{ for } i \geqslant 0)$$

$$\equiv 2^{a_1+\cdots+a_k}.$$

CASE 2. Some $a_i = 0$, say $a_1 = 0$. Now

$$(1+5m)! = (1+5m)(5m)!,$$

so $f(1+5m) \equiv (1+5m)f(5m)$. But $f(5m)$ is even for $m \geqslant 1$ since $(5m)!$ is divisible by a higher power of 2 than of 5. But

$$(1+5m)\cdot(2j) \equiv 2j,$$

so $f(1+5m) \equiv f(5m)$. Letting $m = 5^{a_2-1} + \cdots + 5^{a_k-1}$, the proof follows by induction.

(b) The least $p \geqslant 1$ for which $2^{s+p} \equiv 2^2$ for all $s \geqslant 1$ is $p = 4$.

B-1.

We have

$$f(n+2) - f(n+1) = (n+2)! = (n+2)(n+1)! = (n+2)[f(n+1) - f(n)].$$

It follows that we can take $P(x) = x + 3$ and $Q(x) = -x - 2$.

B-2.

The problem asks for the minimum distance between the quarter of the circle $x^2 + y^2 = 2$ in the open first quadrant and the half of the hyperbola $xy = 9$ in that quadrant. Since the tangents to the respective curves at $(1,1)$ and $(3,3)$ separate the curves and are both perpendicular to $x = y$, the minimum distance is 8.

B-3.

The statement is true. Let ϕ be any bijection on F with no fixed points, and set $x * y = \phi(x)$.

B-4.

Such a function must satisfy

$$\frac{\int_0^x \frac{1}{2} g^2(t)\, dt}{\int_0^x g(t)\, dt} = \frac{1}{x}\int_0^x g(t)\, dt,$$

or equivalently,

$$\int_0^x \frac{1}{2} g^2(t)\, dt = \frac{1}{x}\left[\int_0^x g(t)\, dt\right]^2.$$

Let $z(x) = \int_0^x g(t)\, dt$. Then $z'(x) = g(x)$ and we have

$$\int_0^x \frac{1}{2}(z')^2\, dt = \frac{z^2}{x}, \qquad x > 0.$$

Differentiating, we have

$$\frac{1}{2}(z')^2 = \frac{x \cdot 2zz' - z^2}{x^2}, \qquad x > 0,$$

$$x^2(z')^2 - 4xzz' + 2z^2 = 0, \qquad x > 0,$$

$$(xz' - r_1 z)(xz' - r_2 z) = 0, \qquad x > 0,$$

where $r_1 = 2 + \sqrt{2}$ and $r_2 = 2 - \sqrt{2}$.

Now x, z', and z are continuous and $z > 0$, so the last equation implies that $xz'/z \equiv r$, where $r = r_1$ or $r = r_2$. Separating variables, we have $z'/z = r/x$ and it follows that

$$\ln z = r \ln x + C_0,$$

or equivalently, $z = C_1 x^r$, $C_1 > 0$. Differentiating, we have $z' = g(x) = Cx^{r-1}$, $C > 0$. But g is continuous on $[0, \infty)$ and therefore we cannot have $r = r_2$ (because $r_2 - 1 = 1 - \sqrt{2} < 0$). Thus

$$g(x) = Cx^{1+\sqrt{2}}, \quad C > 0,$$

and one can check that such $g(x)$ do satisfy all the conditions of the problem.

B-5.

Define

$$D(x) = (1 - x)(1 - x^2)(1 - x^4) \cdots (1 - x^{2^{n-1}}).$$

Since binary expansions are unique, each monomial x^k $(0 \leqslant k \leqslant 2^n - 1)$ appears exactly once in the expansion of $D(x)$, with coefficient $(-1)^{d(k)}$. That is,

$$D(x) = \sum_{k=0}^{2^n-1} (-1)^{d(k)} x^k.$$

Applying the operator $\left(x\frac{d}{dx}\right)$ to $D(x)$ m times, we obtain

$$\left(x\frac{d}{dx}\right)^m D(x) = \sum_{k=0}^{2^n-1} (-1)^{d(k)} k^m x^k,$$

so that

$$\left(x\frac{d}{dx}\right)^m D(x)\Big]_{x=1} = \sum_{k=0}^{2^n-1} (-1)^{d(k)} k^m.$$

Define $F(x) = D(x + 1)$, so that

$$\left(x\frac{d}{dx}\right)^m D(x)\Big]_{x=1} = \left[(x+1)\frac{d}{dx}\right]^m F(x)\Big]_{x=0}.$$

But

$$F(x) = \prod_{\alpha=1}^{m} \left[1 - (x+1)^{2^{\alpha-1}}\right] = \prod_{\alpha=1}^{m} \left[-2^{\alpha-1}x + O(x^2)\right], \quad (x \to 0),$$
$$= (-1)^m 2^{m(m-1)/2} x^m + O(x^{m+1}),$$

and by observing that $[(x+1)d/dx]x^n = nx^n + nx^{n-1}$, we see that

$$\left[(x+1)\frac{d}{dx}\right]^m (Ax^m + O(x^{m+1})) = m!A + O(x).$$

So

$$\left[(x+1)\frac{d}{dx}\right]^m F(x)\Big]_{x=0} = (-1)^m 2^{m(m-1)/2} m! + O(x)\Big]_{x=0} = (-1)^m 2^{m(m-1)/2} m!.$$

B-6.

Suppose that $\vec{u}$ and $\vec{v}$ are consecutive edges in P_n. Then $\vec{u}/3$, $(\vec{u}+\vec{v})/3$, and $\vec{v}/3$ are consecutive edges in P_{n+1}. Further,

$$\frac{1}{2}\left\|\frac{\vec{u}}{3}\times\frac{\vec{v}}{3}\right\| = \frac{1}{18}\|\vec{u}\times\vec{v}\|$$

is removed at this corner in making P_{n+1}. But at the next step, the amount from these three consecutive edges is

$$\frac{1}{2}\left\|\frac{\vec{u}}{9}\times\frac{\vec{u}+\vec{v}}{9}\right\| + \frac{1}{2}\left\|\frac{\vec{u}+\vec{v}}{9}\times\frac{\vec{v}}{9}\right\| = \frac{1}{81}\|\vec{u}\times\vec{v}\|.$$

Thus, the amount removed in the $(k+1)$st snip is $2/9$ times the amount removed in the kth.

Note that one-third of the original area is removed at the first step. Thus, the amount removed altogether is

$$\frac{1}{3}\left[1+(2/9)+(2/9)^2+\cdots\right] = \frac{1}{3}\cdot\frac{9}{7} = \frac{3}{7}$$

of the original area. Since the original area is $\sqrt{3}/4$, we have

$$\lim_{n\to\infty}\text{Area } P_n = \frac{4}{7}\cdot\frac{\sqrt{3}}{4} = \frac{\sqrt{3}}{7}.$$

The curve in this problem has been studied extensively by Georges de Rham. (See "Un peu de mathématiques à propos d'une courbe plane," *Elem. Math.*, 2 (1947), 73–76, 89–97; "Sur une courbe plane," *J. Math. Pures Appl.*, 35 (1956), 25–42; and "Sur les courbes limites de polygones obtenus par trisection," *Enseign. Math.*, 5 (1959), 29–43.) Among de Rham's results are the following. The limiting curve is C^1 with zero curvature almost everywhere, but every subarc contains points where the curvature is infinite. Consequently, the curve is nowhere analytic. De Rham parametrizes pieces of the curve so that the tangent vector is intimately related to the Minkowski ?-function. If the construction is repeated, but with each edge divided in the ratio (1/4, 1/2, 1/4) rather than (1/3, 1/3, 1/3), then the resulting limit curve *is* analytic, consisting of piecewise parabolic arcs.

APPENDIX

WINNING TEAMS

Twenty-sixth Competition—1965

Harvard University, Cambridge, Massachusetts
Massachusetts Institute of Technology, Cambridge, Massachusetts
University of Toronto, Toronto, Ontario, Canada
Princeton University, Princeton, New Jersey
California Institute of Technology, Pasadena, California

Twenty-seventh Competition—1966

Harvard University, Cambridge, Massachusetts
Massachusetts Institute of Technology, Cambridge, Massachusetts
University of Chicago, Chicago, Illinois
University of Michigan, Ann Arbor, Michigan
Princeton University, Princeton, New Jersey

Twenty-eighth Competition—1967

Michigan State University, East Lansing, Michigan
California Institute of Technology, Pasadena, California
Harvard University, Cambridge, Massachusetts
Massachusetts Institute of Technology, Cambridge, Massachusetts
University of Michigan, Ann Arbor, Michigan

Twenty-ninth Competition—1968

Massachusetts Institute of Technology, Cambridge, Massachusetts
University of Waterloo, Waterloo, Ontario, Canada
University of California at Los Angeles, Los Angeles, California
Michigan State University, East Lansing, Michigan
University of Kansas, Lawrence, Kansas

Thirtieth Competition—1969

Massachusetts Institute of Technology, Cambridge, Massachusetts
Rice University, Houston, Texas
University of Chicago, Chicago, Illinois
Harvard University, Cambridge, Massachusetts
Yale University, New Haven, Connecticut

Thirty-first Competition–1970

University of Chicago, Chicago, Illinois
Massachusetts Institute of Technology, Cambridge, Massachusetts
University of Toronto, Toronto, Ontario, Canada
Illinois Institute of Technology, Chicago, Illinois
California Institute of Technology, Pasadena, California

Thirty-second Competition—1971

California Institute of Technology, Pasadena, California
University of Chicago, Chicago, Illinois
Harvard University, Cambridge, Massachusetts
University of California, Davis, California
Massachusetts Institute of Technology, Cambridge, Massachusetts

Thirty-third Competition—1972

California Institute of Technology, Pasadena, California
Oberlin College, Oberlin, Ohio
Harvard University, Cambridge, Massachusetts
Swarthmore College, Swarthmore, Pennsylvania
Massachusetts Institute of Technology, Cambridge, Massachusetts

Thirty-fourth Competition—1973

California Institute of Technology, Pasadena, California
University of British Columbia, Vancouver, British Columbia, Canada
University of Chicago, Chicago, Illinois
Harvard University, Cambridge, Massachusetts
Princeton University, Princeton, New Jersey

Thirty-fifth Competition—1974

University of Waterloo, Waterloo, Ontario, Canada
University of Chicago, Chicago, Illinois
California Institute of Technology, Pasadena, California
Massachusetts Institute of Technology, Cambridge, Massachusetts
University of British Columbia, Vancouver, British Columbia, Canada

Thirty-sixth Competition—1975

California Institute of Technology, Pasadena, California
University of Chicago, Chicago, Illinois
Massachusetts Institute of Technology, Cambridge, Massachusetts
Princeton University, Princeton, New Jersey
Harvard University, Cambridge, Massachusetts

Thirty-seventh Competition—1976

California Institute of Technology, Pasadena, California
Washington University, St. Louis, Missouri
Princeton University, Princeton, New Jersey
*Case Western Reserve University, Cleveland, Ohio
*Massachusetts Institute of Technology, Cambridge, Massachusetts

---

*Tied for fourth place.

Thirty-eighth Competition—1977

Washington University, St. Louis, Missouri
University of California, Davis, California
California Institute of Technology, Pasadena, California
Princeton University, Princeton, New Jersey
Massachusetts Institute of Technology, Cambridge, Massachusetts

Thirty-ninth Competition—1978

Case Western Reserve University, Cleveland, Ohio
Washington University, St. Louis, Missouri
University of Waterloo, Waterloo, Ontario, Canada
Harvard University, Cambridge, Massachusetts
California Institute of Technology, Pasadena, California

Fortieth Competition—1979

Massachusetts Institute of Technology, Cambridge, Massachusetts
California Institute of Technology, Pasadena, California
Princeton University, Princeton, New Jersey
Stanford University, Stanford, California
University of Waterloo, Waterloo, Ontario, Canada

Forty-first Competition—1980

Washington University, St. Louis, Missouri
Harvard University, Cambridge, Massachusetts
University of Maryland, College Park, Maryland
University of Chicago, Chicago, Illinois
University of California, Berkeley, California

Forty-second Competition—1981

Washington University, St. Louis, Missouri
Princeton University, Princeton, New Jersey
Harvard University, Cambridge, Massachusetts
Stanford University, Stanford, California
University of Maryland, College Park, Maryland

Forty-third Competition—1982

Harvard University, Cambridge, Massachusetts
University of Waterloo, Waterloo, Ontario, Canada
California Institute of Technology, Pasadena, California
Yale University, New Haven, Connecticut
Princeton University, Princeton, New Jersey

Forty-fourth Competition—1983

California Institute of Technology, Pasadena, California
Washington University, St. Louis, Missouri
University of Waterloo, Waterloo, Ontario, Canada

Princeton University, Princeton, New Jersey
University of Chicago, Chicago, Illinois

Forty-fifth Competition—1984

*University of California, Davis, California
*Washington University, St. Louis, Missouri
Harvard University, Cambridge, Massachusetts
Princeton University, Princeton, New Jersey
Yale University, New Haven, Connecticut

---

* Tied for first place.

WINNING INDIVIDUALS

Twenty-sixth Competition—1965

Andreas R. Blass, University of Detroit
Robert Bowen, University of California, Berkeley
Daniel Fendel, Harvard University
Lon M. Rosen, University of Toronto
Barry Simon, Harvard University

Twenty-seventh Competition—1966

Marshall W. Buck, Harvard University
Theodore C. Chang, Massachusetts Institute of Technology
Robert E. Maas, University of Santa Clara
Richard C. Schroeppel, Massachusetts Institute of Technology
Robert S. Winternitz, Massachusetts Institute of Technology

Twenty-eighth Competition—1967

David R. Haynor, Harvard University
Dennis A. Hejhal, University of Chicago
Peter L. Montgomery, University of California, Berkeley
Richard C. Shroeppel, Massachusetts Institute of Technology
Don B. Zagier, Massachusetts Institute of Technology

Twenty-ninth Competition—1968

Don Coppersmith, Massachusetts Institute of Technology
Gerald A. Edgar, University of California, Santa Barbara
Gerald S. Gras, Massachusetts Institute of Technology
Dean G. Huffman, Yale University
Neal Koblitz, Harvard University

Thirtieth Competition—1969

Alan R. Beale, Rice University
Don Coppersmith, Massachusetts Institute of Technology
Gerald A. Edgar, University of California, Santa Barbara
Robert A. Oliver, University of Chicago
Steven Winkler, Massachusetts Institute of Technology

Thirty-first Competition—1970

Jockum Aniansson, Yale University
Don Coppersmith, Massachusetts Institute of Technology
Jeffrey Lagarias, Massachusetts Institute of Technology
Robert A. Oliver, University of Chicago
Arthur Rubin, Purdue University
Steven K. Winkler, Massachusetts Institute of Technology

Thirty-second Competition—1971

Don Coppersmith, Massachusetts Institute of Technology
Robert Israel, University of Chicago
Dale Peterson, Yale University
Arthur Rubin, Purdue University
David Shucker, Swarthmore College
Michael Yoder, California Institute of Technology

Thirty-third Competition–1972

Ira Gessel, Harvard University
Dean Hickerson, University of California, Davis
Arthur Rothstein, Reed College
Arthur Rubin, California Institute of Technology
David Vogan, University of Chicago
Michael Yoder, California Institute of Technology

Thirty-fourth Competition—1973

David J. Anick, Massachusetts Institute of Technology
Peter G. de Buda, University of Toronto
Matthew L. Ginsberg, Wesleyan University
Arthur L. Rubin, California Institute of Technology
Angelos J. Tsirimokos, Princeton University

Thirty-fifth Competition—1974

Thomas G. Goodwillie, Harvard University
Grant M. Roberts, University of Waterloo
Karl C. Rubin, Princeton University
James B. Saxe, Union College
Philip N. Strenski, Armstrong State College

Thirty-sixth Competition—1975

Franklin T. Adams, University of Chicago
David J. Anick, Massachusetts Institute of Technology
Ernest S. Davis, Massachusetts Institute of Technology
Thomas G. Goodwillie, Harvard University
Christopher L. Henley, California Institute of Technology

Thirty-seventh Competition—1976

Philip I. Harrington, Washington University, St. Louis
Christopher L. Henley, California Institute of Technology
Paul M. Herdig, Case Western Reserve University
Nathaniel S. Kuhn, Harvard University
Steven T. Tschantz, University of California, Berkeley
David J. Wright, Cornell University

Thirty-eighth Competition—1977

Russell D. Lyons, Case Western Reserve University
Stephen W. Modzelewski, Harvard University
Michael Roberts, Massachusetts Institute of Technology
Adam L. Stephanides, University of Chicago
Paul A. Vojta, University of Minnesota, Minneapolis

Thirty-ninth Competition—1978

Randall L. Dougherty, University of California, Berkeley
Mark R. Kleiman, Princeton University
Russell D. Lyons, Case Western Reserve University
Peter W. Shor, California Institute of Technology
Steven T. Tschantz, University of California, Berkeley

Fortieth Competition—1979

Randall L. Dougherty, University of California, Berkeley
Richard Mifflin, Rice University
Mark G. Pleszkoch, University of Virginia
Miller Puckette, Massachusetts Institute of Technology
Charles H. Walter, Princeton University

Forty-first Competition—1980

Eric D. Carlson, Michigan State University
Randall L. Dougherty, University of California, Berkeley
Daniel J. Goldstein, University of Chicago
Laurence E. Penn, Harvard University
Michael Raship, Harvard University

Forty-second Competition—1981

David W. Ash, University of Waterloo
Scott R. Fluhrer, Case Western Reserve University
Michael J. Larsen, Harvard University
Robin A. Pemantle, University of California, Berkeley
Adam Stephanides, University of Chicago

Forty-third Competition—1982

David W. Ash, University of Waterloo
Eric D. Carlson, Michigan State University
Noam D. Elkies, Columbia University
Brian R. Hunt, University of Maryland, College Park
Edward A. Shpiz, Washington University, St. Louis

Forty-fourth Competition—1983

David W. Ash, University of Waterloo
Eric D. Carlson, Michigan State University

Noam D. Elkies, Columbia University
Michael J. Larsen, Harvard University
Gregg N. Patruno, Princeton University

Forty-fifth Competition—1984

Noam D. Elkies, Columbia University
Benji N. Fisher, Harvard University
Daniel W. Johnson, Rose-Hulman Institute of Technology
Michael Reid, Harvard University
Richard A. Stong, Washington University, St. Louis

INDEX OF PROBLEMS